Springer Theses

Recognizing Outstanding Ph.D. Research

For further volumes:
http://www.springer.com/series/8790

Aims and Scope

The series "Springer Theses" brings together a selection of the very best Ph.D. theses from around the world and across the physical sciences. Nominated and endorsed by two recognized specialists, each published volume has been selected for its scientific excellence and the high impact of its contents for the pertinent field of research. For greater accessibility to non-specialists, the published versions include an extended introduction, as well as a foreword by the student's supervisor explaining the special relevance of the work for the field. As a whole, the series will provide a valuable resource both for newcomers to the research fields described, and for other scientists seeking detailed background information on special questions. Finally, it provides an accredited documentation of the valuable contributions made by today's younger generation of scientists.

Theses are accepted into the series by invited nomination only and must fulfill all of the following criteria

- They must be written in good English.
- The topic should fall within the confines of Chemistry, Physics, Earth Sciences, Engineering and related interdisciplinary fields such as Materials, Nanoscience, Chemical Engineering, Complex Systems and Biophysics.
- The work reported in the thesis must represent a significant scientific advance.
- If the thesis includes previously published material, permission to reproduce this must be gained from the respective copyright holder.
- They must have been examined and passed during the 12 months prior to nomination.
- Each thesis should include a foreword by the supervisor outlining the significance of its content.
- The theses should have a clearly defined structure including an introduction accessible to scientists not expert in that particular field.

Jana Nováková

Standard Model Measurements with the ATLAS Detector

Monte Carlo Simulations of the Tile Calorimeter and Measurement of the $Z \to \tau\tau$ Cross Section

Doctoral Thesis accepted by
the Charles University in Prague, Czech Republic

Author
Dr. Jana Nováková
Faculty of Mathematics and Physics
Charles University in Prague
Prague
Czech Republic

Supervisor
Dr. Tomáš Davídek
Faculty of Mathematics and Physics
Charles University in Prague
Prague
Czech Republic

ISSN 2190-5053 ISSN 2190-5061 (electronic)
ISBN 978-3-319-00809-7 ISBN 978-3-319-00810-3 (eBook)
DOI 10.1007/978-3-319-00810-3
Springer Cham Heidelberg New York Dordrecht London

Library of Congress Control Number: 2013942990

© Springer International Publishing Switzerland 2014
This work is subject to copyright. All rights are reserved by the Publisher, whether the whole or part of the material is concerned, specifically the rights of translation, reprinting, reuse of illustrations, recitation, broadcasting, reproduction on microfilms or in any other physical way, and transmission or information storage and retrieval, electronic adaptation, computer software, or by similar or dissimilar methodology now known or hereafter developed. Exempted from this legal reservation are brief excerpts in connection with reviews or scholarly analysis or material supplied specifically for the purpose of being entered and executed on a computer system, for exclusive use by the purchaser of the work. Duplication of this publication or parts thereof is permitted only under the provisions of the Copyright Law of the Publisher's location, in its current version, and permission for use must always be obtained from Springer. Permissions for use may be obtained through RightsLink at the Copyright Clearance Center. Violations are liable to prosecution under the respective Copyright Law. The use of general descriptive names, registered names, trademarks, service marks, etc. in this publication does not imply, even in the absence of a specific statement, that such names are exempt from the relevant protective laws and regulations and therefore free for general use.
While the advice and information in this book are believed to be true and accurate at the date of publication, neither the authors nor the editors nor the publisher can accept any legal responsibility for any errors or omissions that may be made. The publisher makes no warranty, express or implied, with respect to the material contained herein.

Printed on acid-free paper

Springer is part of Springer Science+Business Media (www.springer.com)

Publications Related to This Thesis

1. "Measurement of the Z to tau tau Cross Section with the ATLAS Detector"

   The ATLAS Collaboration

   Phys. Rev. D84 (2011) 112006

2. "Readiness of the ATLAS Tile Calorimeter for LHC collisions"

   The ATLAS Collaboration

   Eur. Phys. J. C (2010) 70

3. "$Z \to \tau\tau$ cross section measurement in proton–proton collisions at 7 TeV with the ATLAS experiment"

   The ATLAS Collaboration

   ATLAS public document ATLAS-CONF-2012-006, CERN, 2012

4. "Description of the Tile Calorimeter Electronic Noise"

   A. Artamonov, L. Fiorini, B. T. Martin, J. Nováková, A. Solodkov, I. Vichou

   ATLAS internal document ATL-TILECAL-INT-2011-002, CERN, 2011

Supervisor's Foreword

The Large Hadron Collider (LHC) at CERN has been built to explore the world of elementary particles and their interactions at an unprecedented high energy and luminosity. The goal is to extend our understanding of the constituents of matter and associated physics processes by precise measurements of features of known particles and their interactions in the frame of the Standard model and, even more importantly, to search for new yet unobserved particles, whose existence would eventually establish an extension of the current theory of elementary particles.

ATLAS is one of the two general-purpose experiments at the LHC. It consists of inner tracking detector immersed in 2 T magnetic field, surrounded by electromagnetic and hadronic calorimeters and a powerful muon spectrometer. The hadronic Tile Calorimeter, made of alternating layers of steel and plastic scintillators, covers the central pseudorapidity region in the ATLAS experiment. Its main tasks are—together with other calorimeters—to provide accurate energy and position measurements of electrons, photons, isolated hadrons, taus, and jets. It also contributes in particle identification and in muon momentum reconstruction.

Jana Nováková started her Ph.D. studies in 2007, still before the LHC began its operation. She joined the Tile Calorimeter group and participated in the detector commissioning. With the advent of LHC collision data she moved to detector operations and, more importantly, she took the leading role in the Tile Calorimeter Monte Carlo simulations. Among other items, she contributed significantly in improving the calorimeter noise description and its modeling in Monte Carlo, which is non-trivial due to specific features of the calorimeter electronics. The proper noise description is critical for the reconstruction of jets as well as for other physics objects' measurements.

Simultaneously, Jana participated in the ATLAS collision data analysis. Her studies of identification and trigger efficiencies of electrons represent important contribution to reconstruction of leptonic τ decays and the cross section measurement of the process $p + p \rightarrow Z \rightarrow \tau + \tau$. These studies become even more important due to the recent discovery of a Higgs-like particle. The mentioned process constitutes the dominant background to the Higgs boson decay to a pair of τ leptons, which is now being extensively studied by the ATLAS collaboration.

The thesis combines two topics, reflecting Jana's work in the Tile Calorimeter group as well as her inputs in the Z boson production cross section measurements using the $Z \rightarrow \tau\tau$ decay channel. Jana often brought genuine ideas in solving problems and she demonstrated how well and accurately things can be done. I appreciated very much being her supervisor.

Prague, April 2013

Tomáš Davídek

Acknowledgments

First of all, I would like to thank Tomáš Davídek, my supervisor, for his help and guidance especially in the work related to the Tile Calorimeter. I appreciate that he was always available for discussions of various problems. He was also very helpful with technical aspects of the work.

I would like to thank my colleagues working at the Tile Calorimeter for helping me in understanding the detector properties. Especially, I would like to express my gratitude to Alexander Solodkov who taught me a lot about the Tile Calorimeter software.

Then I wish to thank the $Z \to \tau\tau$ analysis team, especially Elias Coniavitis who was the leader of the effort and who took care about the machinery to make results of our analysis public.

I am very grateful to Rupert Leitner who continuously supported my work and was available for discussions.

Last but not least, I would like to thank my family for their support and patience.

I acknowledge the financial support by grants of Ministry of Education, Youth and Sports under the reference MSM0021620859 ("Research plan") and LA08032 ("INGO Atlas").

Contents

1 Introduction 1
   References 2

2 Overview of the ATLAS Experiment at the LHC 3
   2.1 Large Hadron Collider 3
   2.2 ATLAS Detector 3
        2.2.1 Coordinate System 4
        2.2.2 Inner Detector 5
        2.2.3 Calorimeters 6
        2.2.4 Muon System 7
        2.2.5 Forward Detectors 7
        2.2.6 Trigger 8
   2.3 Physics Programme at the ATLAS Experiment 9
        2.3.1 Theoretical Introduction 9
        2.3.2 Physics Measurements at the ATLAS Experiment 10
   References 12

3 Monte Carlo Simulations of the Tile Calorimeter 15
   3.1 Overview of the Tile Calorimeter 15
        3.1.1 TileCal Readout System 16
        3.1.2 TileCal Signal Reconstruction 17
        3.1.3 Monte Carlo Simulations 19
   3.2 Electronic Noise 21
        3.2.1 Electronic Noise in Data 22
        3.2.2 Electronic Noise in Monte Carlo Simulations 23
   3.3 Multiple Interactions: Pile-up 24
        3.3.1 Pile-up Classification 25
        3.3.2 Monte Carlo Simulations with Pile-up 26
        3.3.3 Pile-up Constants 27
   References 29

4 Electron Efficiency Measurement 31
  4.1 Electron Reconstruction and Identification 31
     4.1.1 Reconstruction 31
     4.1.2 Identification 32
  4.2 Methodology for the Electron Efficiency Measurements....... 33
  4.3 Electron Identification Efficiency with W Tag
     and Probe Method 34
     4.3.1 Data and Monte Carlo Samples 35
     4.3.2 Event Selection................................ 35
     4.3.3 Background Subtraction......................... 38
     4.3.4 Efficiency and Scale Factors Measurement 41
     4.3.5 Systematic Uncertainties 43
     4.3.6 Application of the Identification Scale Factors
          in Physics Analysis........................... 45
  4.4 Single Electron Trigger Efficiency with W Tag
     and Probe Method 46
  4.5 Electron Isolation Efficiency with Z Tag and Probe Method.... 48
  References 50

5 $Z \rightarrow \tau\tau$ Cross Section Measurement 51
  5.1 Introduction 51
  5.2 Signal and Background Processes 52
     5.2.1 $Z \rightarrow \tau\tau$ Signal Signature..................... 53
     5.2.2 Background Processes....................... 53
     5.2.3 Data and Monte Carlo Samples 54
  5.3 Object and Event Selection............................ 55
     5.3.1 Event Preselection........................... 56
     5.3.2 Object Preselection and Selection 57
     5.3.3 Event Selection in the $\tau_\ell\tau_h$ Channel 61
  5.4 Background Estimation 68
     5.4.1 W+jets Background Normalisation................ 68
     5.4.2 Z+jets Background Normalisation 69
     5.4.3 Multijet Background 70
     5.4.4 Expected Number of Signal and Background Events.... 71
  5.5 Methodology for Cross Section Calculation 71
  5.6 Systematics... 73
     5.6.1 Uncertainties Associated with Electrons............ 74
     5.6.2 Uncertainties on Hadronic τ Candidates 75
     5.6.3 Energy Scale Uncertainty...................... 76
     5.6.4 Background Estimation 77
     5.6.5 Acceptance Factor A_Z Uncertainty 78
     5.6.6 Other Sources of Systematic Uncertainty........... 79
     5.6.7 Summary of the Systematics 79

Contents

5.7   Final Results 80

References 82

6   Conclusions 85

Reference 86

Appendix A: Single Electron Trigger Scale Factors 87

Appendix B: Electron Isolation Scale Factors 89

Abstract

Monte Carlo simulations of the Tile Calorimeter in the ATLAS experiment at the Large Hadron Collider (LHC), especially the electronic noise and multiple interactions (so-called pile-up), are discussed in the thesis. A good agreement in the cell energy distribution between data and Monte Carlo simulations is found. The cross section measurement of $Z \to \tau\tau$ events with the invariant mass between 66 and 116 GeV with the ATLAS experiment is described in the next part of the thesis. Data samples collected during 2011 corresponding to the integrated luminosity of $1.34 - 1.55$ fb^{-1} are used for the analysis. The measurements are performed in three different final states depending on the decay mode of the τ leptons. The measurement in the channel with one τ lepton decaying leptonically into the electron + neutrinos (schematically $\tau \to e + \nu_e + \nu_\tau$) and the other one hadronically (schematically $\tau \to$ hadrons $+ \nu_\tau$), especially the calculation of the nominal cross section and the evaluation of the systematic uncertainties, is discussed in details in the thesis.

Chapter 1
Introduction

The Large Hadron Collider (LHC) is a proton-proton collider operating at CERN. One of the multi-purpose apparatus built at LHC is the ATLAS detector which is designed to allow studies of the widest possible range of physics processes. The thesis is based on the data collected with the ATLAS experiment and on the Monte Carlo simulations of the detector. The analysis presented in this thesis might be divided into two parts. The first part is more technical and the Tile Calorimeter, especially its Monte Carlo simulations, is discussed there. The electron performance and the measurement of the $Z \rightarrow \tau\tau$ cross section are presented in the other part of the thesis.

The Tile Calorimeter is a sampling calorimeter designed primarily for the detection of the hadronic showers created by jets. The treatment of the electronic noise and multiple interactions (so-called pile-up) in Monte Carlo simulations are studied in more details in the thesis. A good description of both noise and pile-up is crucial for the creation of clusters of calorimetric cells that build seeds for jets and hadronic τ jets that are used in many physics analyses performed at ATLAS.

The method used for the measurement of the $Z \rightarrow \tau\tau$ cross section is described in the second part of the thesis. Due to the fact that the τ lepton has a very short lifetime (mean lifetime $c\tau = 87\,\mu\text{m}$ [1]), it cannot be directly seen in the detector, but its decay products are detected. The τ leptons decay leptonically in 35.3 % ($\tau \rightarrow e\nu_e\nu_\tau$, $\tau \rightarrow \mu\nu_\mu\nu_\tau$) or hadronically in 64.7 % (mostly into one or three charged hadrons accompanied with a τ neutrino and possibly also few additional neutral hadrons). Three final states were studied in ATLAS with a data sample collected during 2011: $Z \rightarrow \tau\tau \rightarrow e\mu + 4\nu$ (denoted as $\tau_e\tau_\mu$), $Z \rightarrow \tau\tau \rightarrow \mu + \text{hadrons} + 3\nu$ (denoted as $\tau_\mu\tau_h$) and $Z \rightarrow \tau\tau \rightarrow e + \text{hadrons} + 3\nu$ (denoted as $\tau_e\tau_h$).[1] The latter final state is discussed in this thesis in details since the author was strongly involved in this analysis. The other final states are mentioned briefly for completeness. The measurement of the $Z \rightarrow \tau\tau$ cross section was performed separately with a data sample collected during 2010 and 2011. The study with 2010 data, where the author participated, was published in Ref. [2]. The new analysis using 2011 data, where the

---

[1] A schematic notation is used throughout the thesis—the charge of the decay products is not denoted, nor the neutrino type, neutrino is not distinguished from the anti-neutrino in this notation.

J. Nováková, *Standard Model Measurements with the ATLAS Detector*,
Springer Theses, DOI: 10.1007/978-3-319-00810-3_1,
© Springer International Publishing Switzerland 2014

1

author belonged to the key analysers, was reviewed within the ATLAS collaboration and made available for general public, see Ref. [3].

Due to the presence of the electron in the final state in the $Z \rightarrow \tau\tau \rightarrow e + \text{hadrons} + 3\nu$ decay, the electron reconstruction, identification as well as a measurement of the electron identification efficiency (by so-called Tag and Probe method) are discussed in more details in the thesis. These measurements are crucial for a good agreement between data and Monte Carlo simulations, therefore also for the cross section measurement.

The thesis is organised as follows: A brief description of the ATLAS detector, its composition and a physics programme, is given in Chap. 2. Details about the Monte Carlo simulations of the Tile Calorimeter are presented in Chap. 3. The electron reconstruction and the efficiency measurements are described in Chap. 4, followed by the cross section measurement of the $Z \rightarrow \tau\tau$ process in Chap. 5.

References

1. Particle Data Group Collaboration, K. Nakamura et al., Review of particle physics. J. Phys. **G37**, 075021 (2010)
2. The ATLAS Collaboration, G. Aad et al., Measurement of the Z to tau tau cross section with the ATLAS detector. Phys. Rev. D **84**, 112006 (2011)
3. The ATLAS Collaboration, G. Aad et al., $Z \rightarrow \tau\tau$ cross section measurement in proton-proton collisions at 7 TeV with the ATLAS experiment, ATL-CONF-2012-006, Feb 2012

Chapter 2
Overview of the ATLAS Experiment at the LHC

A brief description of the Large Hadron Collider (LHC) and the experiment ATLAS is presented in this chapter. The LHC design overview (Sect. 2.1) is followed by the short description of the ATLAS detector components (Sect. 2.2). Next, a brief summary of the particle physics theory is given and the main goals of the physics programme at the ATLAS experiment are mentioned (Sect. 2.3).

2.1 Large Hadron Collider

The LHC [1] at CERN was built to allow studies of the particle physics processes at high energy and luminosity conditions that have not been reached before. It was designed to collide proton beams at the centre of mass energy 14 TeV at a luminosity of 10^{34} cm^{-2} s^{-1}. These conditions have not been reached by the end of 2011. After tests at lower energy, the LHC started its operation at 7 TeV centre of mass energy in spring 2010 and continued with the same energy during 2011. Total integrated luminosity delivered to the ATLAS experiment was 0.05 fb^{-1} in year 2010 and 5.61 fb^{-1} in 2011. The peak luminosity reached in ATLAS during 2011 running was 3.65×10^{33} cm^{-2} s^{-1}.

Two general purpose detectors were built to explore the proton-proton and heavy ions collisions at the LHC—ATLAS (A Toroidal LHC ApparatuS) and CMS (Compact Muon Solenoid). Since the measurements described in the thesis have been performed with the ATLAS detector, the ATLAS experiment is described in more details below.

2.2 ATLAS Detector

The design of the ATLAS detector was devised to allow a study of as wide range of physics processes as possible. Particularly, searches for so-far unobserved particles represent an experimental challenge and they define requirements for the ATLAS apparatus.

J. Nováková, *Standard Model Measurements with the ATLAS Detector*,
Springer Theses, DOI: 10.1007/978-3-319-00810-3_2,
© Springer International Publishing Switzerland 2014

2 Overview of the ATLAS Experiment at the LHC

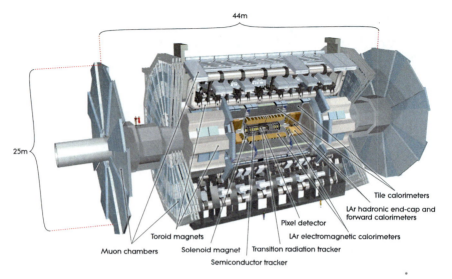

Fig. 2.1 Layout of the ATLAS detector [2]

The ATLAS detector and its expected performance are described in details in Ref. [2]. The detector contains several layers of different sub-detectors, namely the inner detector closest to the beam pipe, electromagnetic and hadronic calorimeters and muon chambers laying in the largest distance from the interaction point. The ATLAS detector design is formed also by magnets. The magnet configuration consists of the superconducting solenoid surrounding the inner detector and three large superconducting toroids (one barrel and two end-caps) arranged with an eight-fold azimuthal symmetry around the calorimeters. The layout of the ATLAS detector is shown in Fig. 2.1.

2.2.1 Coordinate System

The coordinate system used in this thesis follows the standard ATLAS definition [2] which is the following: The nominal interaction point is defined as the origin of the coordinate system, while the beam direction defines the z-axis and the x–y plane is transverse to the beam direction. The positive x-axis is defined as pointing from the interaction point to the centre of the LHC ring and the positive y-axis is defined as pointing upwards. The azimuthal angle ϕ is measured around the beam axis and the polar angle θ is the angle from the beam axis. The pseudorapidity is defined as $\eta = -\ln\tan(\theta/2)$. The transverse momentum p_T, the transverse energy E_T and the missing transverse energy E_T^{miss} are defined in the x–y plane.

The distance ΔR in the $\eta - \phi$ space is defined as $\Delta R = \sqrt{\Delta\eta^2 + \Delta\phi^2}$.

2.2.2 Inner Detector

The tracking detector is placed inside a 2 T magnetic field which is generated by the solenoid magnet surrounding the inner detector. It is designed to provide detailed information about charged particles, namely the transverse momentum and primary and secondary vertices are measured with the tracking detector. The inner detector is built of a number of individual components as shown in Fig. 2.2.

The precision pixel and silicon microstrip (SCT) trackers with a very fine segmentation cover the pseudorapidity range up to $|\eta| < 2.5$. The precision tracking detectors are arranged on concentric cylinders around the beam axis while in the end-caps they are located on disks perpendicular to the beam axis. The first layer of the pixel detector with highest granularity, so-called B-layer, is very important for an excellent vertexing. Typically three pixel layers and eight SCT layers are crossed by a good quality track.

A large number of hits, typically 36 per track, is measured with straw tubes of the Transition Radiation Tracker (TRT) which covers the pseudorapidity region up to $|\eta| < 2.0$ and creates the outermost part of the tracking detector. The TRT detector enables also the electron vs. pion identification through the detection of transition radiation photons in the xenon-based gas mixture of its straw tubes.

The tracking system has an expected resolution of $\sigma_{p_T}/p_T = 0.05\,\% \, p_T \oplus 1\,\%$ (with p_T in GeV) in the whole pseudorapidity coverage $|\eta| < 2.5$ [2].

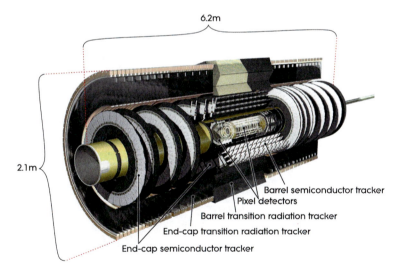

Fig. 2.2 Components of the ATLAS inner detector [2]

2.2.3 Calorimeters

The ATLAS calorimetry system, shown in Fig. 2.3, consists of different types of sampling calorimeters covering the total pseudorapidity range $|\eta| < 4.9$. The fine granularity of the electromagnetic calorimeter in the region matched to the inner detector is necessary for precision measurements of electrons and photons. The hadronic calorimeters are dedicated for the jet reconstruction and missing transverse energy measurement for which a coarser granularity is sufficient.

The electromagnetic (EM) system consists of two parts—a presampler and an EM calorimeter. The EM calorimeter with the liquid argon (LAr) as an active material has a typical structure of an accordion-geometry with kapton electrodes and lead absorber plates. The calorimeter is symmetric in the azimuthal angle without any azimuthal cracks. The calorimeter is built of three longitudinal layers. Most of the EM shower energy for high E_T electrons and photons is collected in the middle layer which has a fine granularity of 0.025×0.025 in $\eta \times \phi$ space. The first layer, so-called strip layer, offers an excellent $\gamma - \pi^0$ discrimination. The last layer with coarser granularity collects the energy deposited in the tail of the very energetic EM showers. The EM calorimeter is divided into a barrel part ($|\eta| < 1.475$) and two end-caps ($1.375 < |\eta| < 3.2$). The presampler detector is located in front of the EM calorimeter in the region $|\eta| < 1.8$. It is developed to correct for the energy lost in the material before the calorimeter. It consists of an active LAr layer of thickness 1.1 cm in the barrel and 0.5 cm in the end-cap.

The Tile Calorimeter is a hadronic calorimeter covering the range $|\eta| < 1.7$ with steel used as an absorber and scintillating tiles as an active material. The design of the

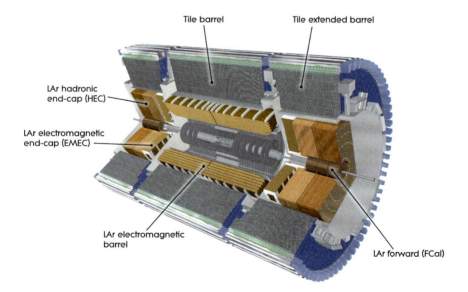

Fig. 2.3 Calorimetric system of the ATLAS detector [2]

2.2 ATLAS Detector

Tile Calorimeter is described in details in Chap. 3. Contrary to the Tile Calorimeter, the forward hadronic calorimeters use the LAr technology. The Hadronic End-cap Calorimeter (HEC) covers pseudorapidity range from 1.5 to 3.2 using copper as the absorber. Finally, the Forward Calorimeter (FCal) covers the most forward region up to $|\eta| < 4.9$. The FCal consists of three modules in each end-cap: The first module is made of copper and is optimised for electromagnetic measurements, the other two are made of tungsten and are used primarily for measurements of the hadronic showers.

The designed resolution of the calorimetric system is the following (with E in GeV) [2]:

- Electromagnetic calorimeter ($|\eta| < 3.2$): $\sigma_E/E = 10\,\%/\sqrt{E} \oplus 0.7\,\%$
- Hadronic calorimeter (jets):

  - Barrel and end-cap ($|\eta| < 3.2$): $\sigma_E/E = 50\,\%/\sqrt{E} \oplus 3\,\%$
  - Forward region ($3.1 < |\eta| < 4.9$): $\sigma_E/E = 100\,\%/\sqrt{E} \oplus 10\,\%$.

2.2.4 Muon System

The muon spectrometer measures the deflection of the muon tracks in the magnetic field produced by large superconduction air-core toroid magnets (one in the barrel and two in the end-caps) in the region $|\eta| < 2.7$. The spectrometer chambers are arranged in three cylindrical layers around the beam axis while in the transition region and in the end-caps the chambers are installed in three planes perpendicular to the beam axis. The layout of the muon chambers is shown in Fig. 2.4.

The Monitored Drift Tubes (MDTs) cover most of the pseudorapidity range of the muon system and provide a precision measurement of the muon tracks. Cathode Strip Chambers (CSCs) with higher granularity are used in the large pseudorapidities ($2.0 < |\eta| < 2.7$). The CSCs are radiation resistant and can be used in a region with an increasing particle rate.

The muon trigger system covers the pseudorapidity range up to 2.4. Resistive Plate Chambers (RPCs) are used in the barrel and Thin Gap Chambers (TGCs) in the end-caps. These chambers are used to measure the muon coordinate in the direction orthogonal to the precision-tracking chambers and also for triggering.

The expected resolution of the muon spectrometer is $\sigma_{p_T}/p_T = 10\,\%$ at $p_T = 1$ TeV [2].

2.2.5 Forward Detectors

The forward region of the pseudorapidity is covered with three smaller detectors—LUCID, ALFA and ZDC.

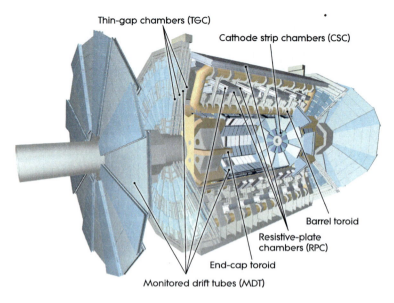

Fig. 2.4 Muon system of the ATLAS detector [2]

The main goal of LUCID (LUminosity measurement using Cerenkov Integrating Detector) and ALFA (Absolute Luminosity For ATLAS) is a measurement of the luminosity delivered to ATLAS. LUCID is located at ±17 m from the interaction point and provides online relative luminosity for ATLAS. ALFA is located at ±240 m and is made of scintillating fibre trackers located inside Roman pots. ALFA is optimised for measuring the absolute value of the delivered luminosity.

The third component of the forward system is called ZDC (Zero-Degree Calorimeter) and is designed to measure the production of neutral particles in the very forward direction ($|\eta| > 8.2$).

2.2.6 Trigger

The collision rate at the design luminosity will be 40 GHz and the final rate of events being saved after the trigger decision might reach 400 Hz maximum. The trigger system in ATLAS consists of three distinct levels: Level 1 (L1), Level 2 (L2) and Event Filter (EF).

The L1 trigger reduces the rate to about 75 kHz and it has to provide the decision within less than 2.5 μs. The L1 trigger searches for high momentum muons, electrons, photons, jets, hadronically decaying τ leptons as well as large missing transverse energy and total transverse energy. It can access only limited information from calorimeters and muon chambers. The L1 trigger defines so-called regions of interest (RoIs)—coordinates in η and ϕ where a high energy object might be located.

2.2 ATLAS Detector

The L2 trigger starts from the RoIs provided by L1. Full granularity and precision is available for these regions at L2 stage. The L2 trigger has the average event processing time of about 40 ms and it is designed to reduce the trigger rate down to 3 kHz.

In the last step, the EF trigger reduces the event rate to approximately 200−400 Hz. The EF can access the full information from the whole detector and it uses some of the offline analysis algorithms within the average event processing time of about 4 s.

2.3 Physics Programme at the ATLAS Experiment

A short description of the particle physics theory is presented, followed by highlights of physics measurements at the ATLAS experiment in this section.

2.3.1 Theoretical Introduction

Interactions between fundamental components of the matter are described by Standard Model which is a quantum field theory based on $SU(3) \times SU(2) \times U(1)$ symmetry group. The interactions between the constituents of the matter (spin 1/2 fermions) are described be an exchange of intermediate bosons with spin 1.

The quantum chromodynamic (QCD) [3] is an $SU(3)$ gauge theory where the colour charge defines the local symmetry. The interactions between quarks, which form a triplet under colour $SU(3)$, are mediated by the exchange of gluons. The eight gluons carry the colour charge and therefore are self-interacting. Unlike the electromagnetic interaction, the strong interaction strength is small for large momentum transfers (asymptotic freedom) and large for small momentum transfers (confinement). The confinement gives an explanation why the quarks form colour neutral hadrons and cannot be observed individually.

The electroweak interactions [4] are described by an $SU(2) \times U(1)$ symmetry group. Left-handed leptons and quarks form doublets under $SU(2)$, while the right-handed states are singlets. Four intermediate bosons appear in the electroweak theory—a photon, Z and $W^{\pm}$ bosons. One of the crucial differences between the photon and the Z and $W^{\pm}$ bosons is that the photon is massless while the weak interaction bosons are very heavy. The mass terms of the intermediate bosons can be introduced in the Standard Model by spontaneous breaking of the $SU(2) \times U(1)$ symmetry, so-called Higgs mechanism. The simplest way is to consider an $SU(2)$ doublet of complex scalar fields where three degrees of freedom are absorbed in the mass terms of the Z and $W^{\pm}$ bosons and the remaining degree of freedom results in the physical state—the Higgs boson. The Higgs mechanism has not been confirmed experimentally yet. It has been one of the main goals of the ATLAS experiment to find the Higgs boson particle if it exists.

Despite the fact that the Standard Model describes well the experimental data, the model has several theoretical shortcomings (e.g. fine tuning of the Standard Model parameters, large number of free parameters, no candidate for dark matter) and therefore it is not expected to be the final model. Based on the excellent agreement with the experiment, the new model must obtain the Standard Model theory in the limit of low energy. One of the possible extensions of the Standard Model is the Supersymmetry [5] where a supersymmetric partner is assigned to each Standard Model particle—bosonic partners (squarks and sleptons) to fermions and fermionic partners (gluoinos and gauginos) to bosons. In the minimal supersymmetric extension of the Standard Model (MSSM) the Higgs boson sector contains five physical states - two scalars (h, H), one pseudoscalar (A) and two charged Higgs bosons ($H^{\pm}$). Apart from Supersymmetry, also other alternative theories to the Standard Model exist, e.g. technicolour. No particle beyond the Standard Model has been observed experimentally yet and it is a challenge for the ATLAS experiment to find a signature of the physics beyond the Standard Model.

2.3.2 Physics Measurements at the ATLAS Experiment

The physics programme at the ATLAS experiment [5] can be divided into two main parts - precision measurements of the Standard Model properties and searches for not yet observed particles (Higgs boson and particles beyond the Standard Model). The Standard Model measurements and the Higgs boson searches are briefly discussed below.

Standard Model Measurements

The Standard Model measurements play an important role for several reasons. First of all, the Standard Model processes represent the main background for the signatures of the new physics. Therefore an excellent understanding of the Standard Model background is very crucial. Moreover, the well-known processes such as decays of W and Z bosons to leptons are used as a standard candle to calibrate the detector and to correct the Monte Carlo simulations to agree with measurements using real data samples (e.g. smearing of the lepton energy/momentum, correcting efficiency predictions in simulations). This process is illustrated in Chap. 4. Next, precision measurements of the Standard Model parameters can be performed at the ATLAS experiment (e.g. triple gauge boson couplings in diboson final states, W boson mass). Finally, deviations from the Standard Model predictions might indicate physics beyond the Standard Model (e.g. deviations in the high-p_T spectrum of dijet events might be a sign of quark compositness).

The top quark physics is an important part of the Standard Model programme at the ATLAS experiment. Precise measurements of the top mass as well as other top

2.3 Physics Programme at the ATLAS Experiment

quark properties (e.g. charge) are performed with the ATLAS detector. The single top production cross section is also being measured.

Besides the high energy physics, also B-physics measurements are performed at ATLAS experiment. The specific B-physics topics include the measurement of CP violation, B_s^0 mixing and a search for rare decays. Study of B hadrons' decays and spectroscopy is also a part of the B-physics programme.

Higgs Boson Searches

The search for the Higgs boson particle is one of the main goals of the ATLAS experiment. The Higgs boson is predicted by the Standard Model, however its mass is a free parameter. Due to theoretical constraints, the Higgs boson mass must be lower than approximately 800 GeV [6]. The direct search at LEP excluded the Higgs boson with the mass lower than 114.5 GeV at 95 % confidence level (CL) [7].

The Higgs boson searches have been performed by ATLAS and CMS experiments at the LHC. Both experiments announced an observation of a new particle which could be the Higgs boson particle in July 2012. The results are published in Ref. [8], resp. [9] for the ATLAS, resp. CMS experiment. The properties of the observed particle are compatible with the Standard Model Higgs boson, but more data are needed to confirm this hypothesis. The observed Higgs-like particle has a mass of approximately 125 GeV.

The most important Higgs boson decay modes used in ATLAS in the mass region of the new Higgs-like particle are:

- $H \rightarrow \gamma\gamma$ Although the branching ratio of this channel is very low (BR ≈ 0.2 %), it provides the best sensitivity in the low mass region. A very good photon identification, robust against multijets that might be possibly faking photons, as well as an excellent energy reconstruction are the key ingredients for the measurement in this channel. The diphoton invariant mass spectrum is used as the discriminating variable. More details about the observation of the Higgs-like boson in this channel can be found in Ref. [10].
- $H \rightarrow ZZ \rightarrow 4\ell$ This channel would give the cleanest signal in the intermediate Higgs boson mass range. However, the branching ratio below 150 GeV decreases rapidly and it is more difficult to use this channel also in the low mass range. Two pairs of isolated leptons with same flavour and opposite charges are searched for. The invariant mass of the four leptons' system is considered as the discriminant. A high efficiency of the lepton trigger and lepton identification as well as a very good energy resolution are required for the precise measurement in this decay mode. Details about the measurement of the properties of the Higgs-like particle in this channel can be found in Ref. [11].
- $H \rightarrow WW \rightarrow \ell\nu\ell\nu$ The decay mode with two W bosons leads to the best sensitivity in the intermediate Higgs boson mass range and it can also contribute in the lowest Higgs boson mass range. The typical signature is formed by a pair of isolated leptons with opposite charges and large missing transverse energy. The

invariant mass cannot be reconstructed and therefore the transverse mass of the Higgs boson is used as the discriminant. The recent analysis in this channel using full 2011 and 2012 datasets is described in Ref. [12].

- $H \rightarrow \tau\tau$ The Higgs decay mode with a pair of τ leptons is measurable in the low mass region. The difficulty of the search in this channel comes from the hadronic τ which might appear in the final state as a jet can be easily misidentified as a hadronic τ. Therefore the large multijet background must be carefully taken under control. The invariant mass of the Higgs boson cannot be directly reconstructed due to neutrinos from the τ leptons' decays. Alternative mass reconstruction technique, so-called Missing Mass Calculator [13], is used in this channel. The $Z \rightarrow \tau\tau$ process, studied in this thesis, forms the dominant background. Thus a good understanding and description of this background is crucial for the $H \rightarrow \tau\tau$ analysis. The Higgs-like particle has not been observed in this channel by the ATLAS experiment yet. It is an important decay mode which could confirm that the Higgs boson decays into a pair of fermions. The analysis using full 2011 data and a part of 2012 data corresponding to 4.7 fb^{-1} + 13.0 fb^{-1} is described in Ref. [14].
- $H \rightarrow b\bar{b}$ This is the decay mode with the largest branching ratio in the low mass region, however very difficult to detect over the overwhelming multijet background. It can be performed only in the associated production of the Higgs boson with electroweak bosons (ZH, WH) or with a top quark pair ($t\bar{t}H$). The Higgs-like particle has not been observed in this channel yet. More details about the analysis can be found in Ref. [15].

References

1. L. Evans, P. Bryant, LHC Machine. JINST **3**, S08001 (2008)
2. The ATLAS Collaboration, G. Aad et al., The ATLAS experiment at the CERN large hadron collider. JINST **3**, S08003 (2008)
3. J. Chýla, *Quarks, partons and quantum chromodynamics*, http://www-hep2.fzu.cz/Theory/text.pdf (2004)
4. J. Hořejší, *Fundamentals of electroweak theory*. Karolinum Press, Prague, Czech Republic (2002)
5. The ATLAS Collaboration, G. Aad et al., ATLAS: detector and physics performance technical design report, vol 2 (1999)
6. W. Marciano, G. Valencia, S. Willenbrock, Renormalization-group-improved unitary bounds on the Higgs-boson and top-quark masses. Phys. Rev. D **40**, 1725 (1989)
7. ALEPH, DELPHI, L3, OPAL and the LEP Working Group for Higgs boson searches, Search for the standard model Higgs boson at LEP. Phys. Lett. **B565**, 61–75 (2003)
8. The ATLAS Collaboration, Observation of a new particle in the search for the standard model Higgs boson with the ATLAS detector at the LHC. Phys. Lett. B **B716**, 1–29 (2012)
9. The CMS Collaboration, Observation of a new boson at a mass of 125 GeV with the CMS experiment at the LHC. Phys. Lett. B **B716**, 30–61 (2012)
10. The ATLAS Collaboration, G. Aad et al., Measurements of the properties of the Higgs-like boson in the two photon decay channel with the ATLAS detector using 25 fb^{-1} of proton-proton collision data, ATL-CONF-2013-012, Mar 2013

References

11. The ATLAS Collaboration, G. Aad et al., Measurements of the properties of the Higgs-like boson in the four lepton decay channel with the ATLAS detector using 25 fb^{-1} of proton-proton collision data, ATL-CONF-2013-013, Mar 2013
12. The ATLAS Collaboration, G. Aad et al., Measurements of the properties of the Higgs-like boson in the $WW^{(*)} \to \ell\nu\ell\nu$ decay channel with the ATLAS detector using 25 fb^{-1} of proton-proton collision data, ATL-CONF-2013-030, Mar 2013
13. A. Elagin, P. Murat, A. Pranko, A. Safonov, A new mass reconstruction technique for resonances decaying to $\tau\tau$. Nucl. Instrum. Methods Phys. Res. A **654**, 481–489 (2011)
14. The ATLAS Collaboration, G. Aad et al., Search for the standard model Higgs boson in $H \to \tau\tau$ decays in proton-proton collisions with the ATLAS detector, ATL-CONF-2012-160, Nov 2012
15. The ATLAS Collaboration, G. Aad et al., Search for the standard model Higgs boson produced in association with a vector boson and decaying to bottom quarks with the ATLAS detector, ATL-CONF-2012-161, Nov 2012

Chapter 3
Monte Carlo Simulations of the Tile Calorimeter

The Tile Calorimeter (TileCal) is the barrel hadronic calorimeter of the ATLAS detector. Its performance and readiness for the LHC collisions based on special calibration runs, detection of cosmic ray muons and single beam events is described in details in Ref.[1].

The Tile Calorimeter overview is given in Sect. 3.1. The Monte Carlo simulations of the Tile Calorimeter are the main topic of this chapter. The emphasis is given on the treatment of the electronic noise (Sect. 3.2) and the description of multiple interactions per bunch crossing, so-called pile-up (Sect. 3.3).

3.1 Overview of the Tile Calorimeter

TileCal is a sampling hadronic calorimeter covering the pseudorapidity region $-1.7 < \eta < 1.7$. It consists of alternating layers of plastic scintillator (active medium) and iron (absorber). The calorimeter is divided into a barrel ($|\eta| < 1.0$) and two extended barrels on both sides ($0.8 < |\eta| < 1.7$). The Tile Calorimeter has an internal structure and both the barrel and extended barrels are segmented into 64 modules corresponding to $\Delta\phi$ granularity of ~ 0.1 radians. Furthermore, each module is segmented in pseudorapidity and radially as well. The radial segmentation divides the module into three parts corresponding to approximately 1.5, 4.1 and 1.8 λ (nuclear interaction lengths for protons) thickness in the barrel and 1.5, 2.6 and 3.3 λ in the extended barrels. The pseudorapidity segmentation corresponds to $\Delta\eta$ granularity of 0.1 in the first two radial layers (layer A and BC) and 0.2 in the last layer (layer D) as shown in Fig. 3.1. The ϕ, η and radial segmentation defines the three dimensional cells in TileCal.

J. Nováková, *Standard Model Measurements with the ATLAS Detector*,
Springer Theses, DOI: 10.1007/978-3-319-00810-3_3,
© Springer International Publishing Switzerland 2014

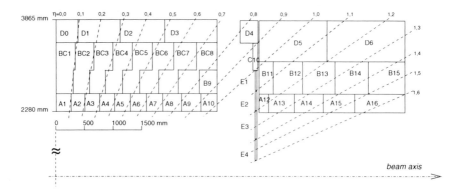

Fig. 3.1 The radial and pseudorapidity segmentation of the TileCal modules in the barrel (*left hand side*) and in the extended barrel (*right hand side*) [1]

3.1.1 TileCal Readout System

Each cell consists of dozens of scintillating tiles and iron plates which are oriented perpendicular to the beam axis and radially staggered [1] as shown in Fig. 3.2. The light produced in the scintillators is collected by the wavelength shifting fibres that are located on both ϕ sides of the modules. The wavelength shifters corresponding to one cell bring the light to two different photomultiplier tubes (PMTs), each on one side of the module. The two PMTs are linked to individual readout channels and the corresponding cell energy is the sum of the energy measured in the two channels. The double readout reduces the dependence on the light attenuation in the scintillators and improves the response uniformity. Furthermore, in case of a single channel problem the information from the other channel is used and the cell energy is twice the energy of the available channel.

The readout electronics (including the PMTs) [1] is housed at the outer radius of the calorimeter. First, the signal from the PMTs is shaped in the way that all pulses have the same width (full width at the half maximum, FWHM, is 50 ns). Thus the energy deposit is proportional to the pulse amplitude. Next, the shaped pulse is amplified in separate high (HG) and low (LG) gain branches with the gain ratio of 64:1. The HG and LG analog signals are sampled with $\Delta t = 25$ ns corresponding to LHC bunch-crossing frequency of 40 MHz. In the case of positive trigger decision, seven samples from one gain for each channel are read out and sent via optical fibres to the backend electronics located outside the experimental hall. The time and energy are determined from the seven samples as described in the next section.

3.1 Overview of the Tile Calorimeter

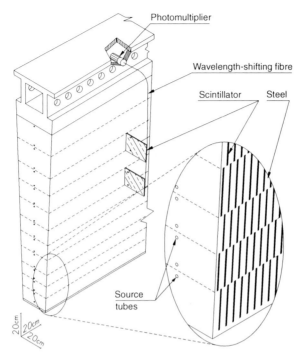

Fig. 3.2 The structure of one of the TileCal modules and the optical readout (the tiles, the fibres, wavelength shifter and PMTs) [1]

3.1.2 TileCal Signal Reconstruction

The pulse amplitude, time and pedestal for each channel are derived by means of the Optimal Filtering (OF) method [2] which is based on the weighted linear combination of the measured samples. The signal amplitude A and time τ are given by equations

$$A = \sum_{i=0}^{n=7} a_i S_i, \quad \tau = \frac{1}{A} \sum_{i=0}^{n=7} b_i S_i \quad (3.1)$$

where S_i is the sample at time t_i ($i = 1, ..., n$) and the coefficients a_i, b_i are derived by OF using the knowledge of the pulse shape and noise autocorrelation matrix. The shape of the pulse for both low gain and high gain read-out is shown in Fig. 3.3.

The timing of each channel with respect to the LHC clock is measured in dedicated calibration runs and with the single beam data. The timing offsets are corrected in the way that the reconstructed time τ is compatible with zero for energy deposits coming from the interaction point and travelling with speed of light.

The OF with fixed time phase, so-called non-iterative OF, is based on the known time phase between the pulse peak and the LHC clock signal for each channel. Due

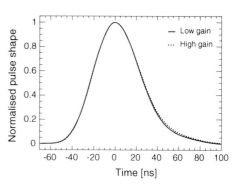

Fig. 3.3 Normalised pulse shape for high and low gain used for the OF weights calculation

to the simplicity of the non-iterative OF algorithm, it can be performed online at the hardware level and the individual samples do not have to be stored for the offline processing. If the individual samples are saved, the iterative OF can be used. The time phase is not fixed in this case and the pulse peak position is searched for iteratively. Both methods, the OF with and without iterations, agree well for the signal-like energy deposits coming from proton-proton collisions. However, they have different performance for the noise-like energy deposits as will be shown in Sect. 3.2.1.

The iterative OF was used as the default reconstruction method during the collisions in 2010 and in the beginning of 2011. The non-iterative OF has become the default reconstruction technique during 2011 data taking. The non-iterative OF is also used for the amplitude reconstruction in the Monte Carlo simulations. Most of the figures in this chapter are obtained with the non-iterative OF reconstruction technique unless explicitly stated otherwise (e.g. data from 2010 are used or the comparison of the two methods is shown).

After the amplitude is reconstructed with OF, the calibration of the channel energy to the electromagnetic scale (EM) is performed

$$E_{\text{channel}} = A \cdot C_{\text{ADC} \to \text{pC}} \cdot C_{\text{Cs}} \cdot C_{\text{pC} \to \text{GeV}} \cdot C_{\text{Laser}} \tag{3.2}$$

where E_{channel} is the channel energy measured in GeV, A is the signal amplitude in ADC counts defined by Eq. (3.1) and the constants C_{XX} represent individual calibration factors. The factor $C_{\text{ADC} \to \text{pC}}$ converting ADC counts to charge in pC is measured for each channel and both gains using a well known injected charge with the Charge Injection System (CIS). The factor C_{Cs} is measured in special calibration runs with Cs radioactive source. First, gain equalisation of all channels is performed with the Cs calibration system. Second, the C_{Cs} factor correcting for residual non-uniformities is derived. The conversion factor $C_{\text{pC} \to \text{GeV}}$ converting charge to energy in GeV was measured in the test beam with electron beams of known energy whose response was analysed. Last calibration system is the laser system which is adapted to

3.1 Overview of the Tile Calorimeter

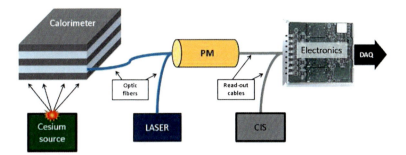

Fig. 3.4 Scheme of the TileCal calibration system

measure and correct for non-linearities of the PMT response.[1] The TileCal calibration system is schematically shown in Fig. 3.4 and more details can be found in Ref. [1].

The cell energy at the EM scale is given by the sum of the reconstructed energy in the two associated channels. If one of the readout channels is masked, the cell energy is twice the energy of the other channel. Thus the energy reconstruction is robust against single channel failures.

3.1.3 Monte Carlo Simulations

Monte Carlo simulations of the TileCal based on the Geant4 programme [3] can be divided into three separate parts: the material description of the detector, simulation of the particle passage through the detector and the signal reconstruction. The signal reconstruction in the Monte Carlo simulations, i.e. the procedure from the hit energy in the scintillator to the cell energy at the EM scale, is described in this section.

The digitisation procedure in the simulations starts from a collection of Geant4 hits in the active material (scintillators) in the TileCal. Each hit is characterised by its energy in MeV, time and position. First, hits belonging to the same channel within one time bin (0.5 ns) are merged together. The sampling fraction correction which converts the scintillator energy to the cell energy is applied in the next step. The energy in MeV is converted to energy in ADC counts using inverse calibration constants defined in Eq. (3.2). Then the hit energy is convoluted with the pulse shape function and seven samples separated by 25 ns are derived. The pedestal, electronic noise and pile-up are added to the individual samples. More details about the electronic noise and pile-up implementation in the Monte Carlo simulations are given in Sects. 3.2 and 3.3. The samples' values are rounded to an integer to correspond to ADC units used in the real data readout procedure. The seven samples represent the input to the signal

---

[1] The laser calibration constants were not used in the calibration loop during 2010 and 2011 data taking, but they are implemented in the energy calibration since 2012 data taking.

reconstruction as described in Sect. 3.1.2. The amplitude reconstruction procedure in the Monte Carlo simulations follows exactly the same scheme as performed in real data.

Sampling Fraction in TileCal

The sampling fraction (SF) correction, the conversion factor between the energy released in the scintillators and the energy deposited in the TileCal cells, depends on the multiple scattering model used in the Monte Carlo simulations, i.e. on the version of the Geant4 programme. It is derived from the TileCal standalone Monte Carlo simulations using the electron beam at defined energy (E_{beam}). If the invisible energy and energy leakage are neglected, the sampling fraction equals

$$SF = \frac{E_{beam}}{E_{scintillator}} \tag{3.3}$$

where $E_{scintillator}$ stands for Geant4 energy deposits in the active medium (scintillators).

Due to the fact that the scintillating tiles are located in planes perpendicular to the LHC beams, the energy released in the scintillators varies with the impact point[4]. Since the electromagnetic showers are very narrow and their dimensions are comparable with the TileCal internal structure, this feature can be observed in the response to the electron beams as shown for the Monte Carlo simulations in Fig. 3.5 (*left hand side*). The dependence of the response on the impact point coordinate z can be described by a simple periodic function

$$E_{scintillator}(z) = p_0 + p_1 \sin(\frac{2\pi}{p_2}z + p_3) \tag{3.4}$$

where p_0 is the energy deposit in the scintillators corrected for the impact point dependence. The variable p_0 is used for the evaluation of the sampling fraction constant instead of $E_{scintillator}$ in Eq. (3.3). The parameter p_1 specifies the relative amplitude of the oscillations, p_2 stands for the thickness of the period as seen by the beam at the given impact angle and p_3 denotes the phase of the oscillations. The numerical values of all four parameters are summarised in Fig. 3.5 (*left hand side*) for the electron beam with energy of 100 GeV at $\eta = 0.35$.

The dependence of the sampling fraction on pseudorapidity is shown in Fig. 3.5 (*right hand side*). The increase of the SF, i.e. decrease of the energy deposit in the scintillators, at small η ($\eta = 0.05$) can be qualitatively explained by the periodic scintillator/iron structure of the TileCal. The area in the scintillators touched by the narrow EM shower is smaller at low angles. The large increase of the SF at large η ($\eta = 0.95$) is caused by the leakage outside the barrel module. On the contrary, the sampling fraction is almost constant in the pseudorapidity region between 0.2 and 0.8.

3.1 Overview of the Tile Calorimeter

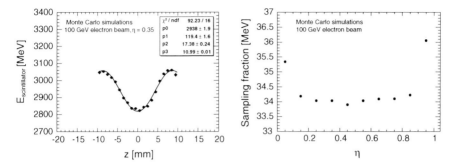

Fig. 3.5 Dependence of the energy deposit in the scintillators on the impact point coordinate z (*left hand side*) and the sampling fraction dependence on pseudorapidity (*right hand side*) are shown. The results are obtained with TileCal standalone simulations with 100 GeV electron beam. The Geant4 programme version 9.4 is used for the simulations

The sampling fraction constant used in the Monte Carlo simulations is the sampling fraction at $\eta = 0.35$ which equals to $SF = 34.0$. The pseudorapidity of 0.35 corresponds to the test beam angle where the electromagnetic scale in TileCal is defined [1].

3.2 Electronic Noise

A good understanding of the electronic noise is crucial for the construction of topological clusters [5] which are constituents for the jets or missing transverse energy calculation. The clustering algorithm searches for energy deposits in the calorimeter significantly above the noise fluctuation level. The algorithm starts from cells with energy above a certain threshold. The default threshold is given by the requirement that the probability of the cell energy to be a noise fluctuation is less than 6.3×10^{-5} corresponding to 4σ for a normal distribution. The seed cluster is expanded to the neighbour cells with energy with the probability less than 4.6×10^{-2} (corresponding to 2σ for a normal distribution) to be a noise contribution. All immediate neighbour cells are added to the cluster in the last step. Thus an unrealistic description of the noise fluctuations in both data and Monte Carlo simulations might affect the shape of the clusters or increase the probability of creating fake clusters.

The noise pattern observed in data is described in Sect. 3.2.1 whereas the implementation of the electronic noise in the Monte Carlo simulations is summarised in Sect. 3.2.2.

3.2.1 Electronic Noise in Data

The electronic noise of the TileCal readout system has been investigated and monitored in special standalone bi-gain runs (so-called pedestal runs) on a long term basis. The pedestal runs are taken regularly during the moments without collisions in the ATLAS detector, typically twice a week.

An example of the typical noise distribution for one cell (cell A9 in module LBA02 corresponding to $\eta = 0.85$ and $\phi = 0.15$ in the inner most layer) is shown in Fig. 3.6. The cell noise was found to be have significant non-Gaussian tail and the shape can be described by a double Gaussian function as shown in the figure. The double Gaussian function (f_{2g}), normalised to $\int f_{2g}(x)dx = 1$, can be generally defined in the following way

$$f_{2g}(x) = \frac{1}{1+R}\left[\frac{1}{\sqrt{2\pi}\sigma_1}\exp\left(-\frac{(x-\mu_1)^2}{2\sigma_1^2}\right) + \frac{R}{\sqrt{2\pi}\sigma_2}\exp\left(-\frac{(x-\mu_2)^2}{2\sigma_2^2}\right)\right] \quad (3.5)$$

where $\sigma_i, \mu_i, i = 1, 2$ are the mean and the width of the two Gaussians and R is the relative share of the two Gaussian functions.

Since parameters μ_i have been found to be negligible as expected for the electronic noise, they have been constrained to $\mu_i = 0$. Thus the noise distribution in TileCal is described by three independent parameters R, σ_1 and σ_2. These parameters are derived cell by cell from data measured in the pedestal runs. The values are stored in the database and are used for the cluster creation to define the 4σ and 2σ limits.

The cell noise averaged over ϕ as a function of pseudorapidity and radial layer is shown in Fig. 3.7. Since it is not straight-forward how to compare all three parameters

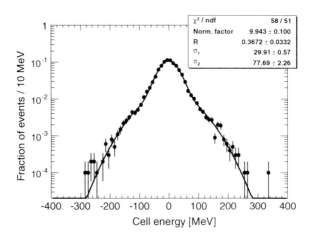

Fig. 3.6 Typical electronic noise distribution in high gain for one cell (cell A9 in module LBA02). The meaning of the constants is given in Eq. (3.5), the variables σ_1 and σ_2 are measured in MeV

3.2 Electronic Noise

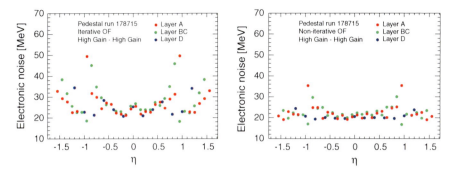

Fig. 3.7 Cell noise averaged over ϕ as a function of η for reconstruction with OF with iterations (*left hand side*) and without iterations (*right hand side*) measured in a pedestal calibration run for the high gain read-out in both channels. A special tool reducing the contribution of the correlated noise for the channels with electronics on the same motherboard has been applied to the data

of the double Gaussian function, the spread[2] of the noise distribution is used as an estimator of the cell noise in this case. The electronic noise varies from 20 to 50 MeV if the iterative OF method is used (*left hand side figure*). The noise values are lower for the non-iterative OF (*right hand side figure*), typical values are between 15 and 35 MeV. It can be explained by the fact that the time phase is fixed for the non-iterative OF and no search for the pulse peak is performed. Moreover, the electronic noise is more uniform across η compared to the iterative OF, especially in the extended barrels. It has been observed that cells whose electronics are located at the outer boundaries of the TileCal barrel and extended barrel modules suffer from higher noise level compared to cells located in the central region.[3]

3.2.2 Electronic Noise in Monte Carlo Simulations

The electronic noise is added to the individual samples in the Monte Carlo simulations as mentioned in Sect. 3.1.3. The basic requirement for the noise implementation in the Monte Carlo is a good agreement of the noise shape in data and in simulations at the cell level. Thus the shape of the noise function used at the digitisation level in simulations has to be such that after the reconstruction of amplitude with OF the cell energy has the double Gaussian shape as observed in data. It has been found that the double Gaussian function fulfils this requirement to a good precision.

---

[2] The spread means the standard deviation (called RMS in ROOT analysis framework [6]), i.e. $\sqrt{(1/N) \cdot (\sum_i (x_i - x_{mean})^2)}$.

[3] It has been found that the electronic noise is largely influenced by the Low Voltage Power Supplies (LVPS). The new generation of the LVPS being currently tested gives promising results with more uniform electronic noise across the η coordinate and almost Gaussian noise distribution.

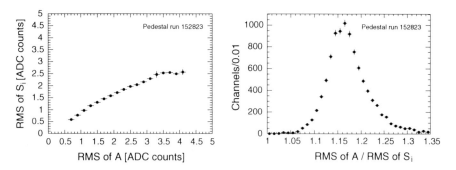

Fig. 3.8 Relation between a spread of reconstructed amplitudes ("RMS of A") and a spread of samples ("RMS of S_i") measured in a pedestal calibration run for the high gain read-out is shown. The non-iterative OF is used for the amplitude reconstruction

Three parameters, R, σ_1 and σ_2, defining the double Gaussian function according to Eq. (3.5) which are to be applied on the individual samples in simulations have to be found for each channel. They are derived from the double Gaussian parameters at the cell level measured in real data. The normalisation factor R is assumed to remain the same before and after the application of the OF, but the values of σ_1 and σ_2 have to be scaled to get the appropriate values at the digital level. The relation between the spread of the samples ("RMS of S_i") and the reconstructed amplitudes with OF ("RMS of A") has been measured in data and is shown in Fig. 3.8. The ratio has been found to be constant over a large range of the noise values. The typical noise values at the sample level are between 1.1 and 1.8 ADC counts in the high gain. The ratio of the spread of reconstructed amplitudes and the spread of the samples ("RMS of A/RMS of S_i") depends on the reconstruction method used. The distribution of this ratio for the non-iterative OF in data is shown on the right hand side in Fig. 3.8 and it can be characterised by a mean value of 1.17 and a spread of 0.05. The mean value is used as the scaling factor from the σ_1 and σ_2 after the OF to the digital level constants which are to be applied to the individual samples in the Monte Carlo simulations.

The comparison between the cell energy distribution in data and in Monte Carlo simulations is shown in Fig. 3.9. The overall agreement is found to be good, even in the low energy region dominated by the electronic noise contribution. However, the noise description in the Monte Carlo simulations does not cover all features of the real electronic noise, namely the correlations between the individual channels are not described in the Monte Carlo simulations.

3.3 Multiple Interactions: Pile-up

The high energy signal reconstruction in the calorimeter is influenced by multiple proton-proton interactions within the bunch crossing (so-called pile-up) at the LHC running conditions. The impact of pile-up on the cell energy distribution is shown in this section.

3.3 Multiple Interactions: Pile-up

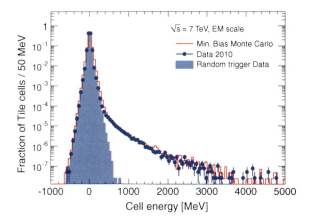

Fig. 3.9 TileCal cells' energy distribution calibrated to the EM scale in data from 2010 and in Monte Carlo simulations. The cell noise in data during the collisions can be estimated in randomly selected events (events passing so-called random trigger). The iterative OF is used for the reconstruction of the cell energy in data and the non-iterative OF in the MC simulations

The cell energy spectra have been studied qualitatively and quantitatively in both data and Monte Carlo simulations. The studies with data from 2011 confirm a reasonably good agreement between Monte Carlo predictions and measurements in data. The author of the thesis was involved in the analysis with the Monte Carlo simulations only. Therefore results based purely on the simulations are shown in this section.

3.3.1 Pile-up Classification

The average number of the minimum bias collisions per bunch crossing depends on the proton beam parameters which vary during the data taking periods at the LHC. According to Ref. [7], the mean number of interactions per crossing (μ) is proportional to the instantaneous luminosity ($\mathcal{L}$) following the relation

$$\mu = \frac{\mathcal{L} \cdot \sigma_{\text{inel}}}{n_{\text{bunch}} \cdot f_r} \qquad (3.6)$$

where σ_{inel} denotes the proton-proton inelastic cross section ($\sigma_{\text{inel}} = 71.5$ mb), n_{bunch} is the number of colliding bunches and f_r is the LHC revolution frequency.

The mean number of the underlying interactions during 2010 and 2011 data taking periods can be found in Ref. [8]. Data collected during 2010 can be characterised by a low number of minimum bias collisions per bunch crossing with μ between approximately 0 and 3 depending on the period. Data taken during 2011 suffer from larger pile-up contribution with the average number of pile-up collisions in the first part of the data taking period of approximately 6 and it increases to the mean value of 12 after the technical stop in september.

The soft interactions, that can affect the energy reconstruction in the calorimeter, come either from the same bunch crossing as the high-energy interaction or from nearby bunch crossings. Two categories of pile-up are defined based on this categorisation:

- **In-time pile-up** stands for multiple interactions coming from the same bunch crossing as the high energy collision. The signal in a given cell might be a sum of the energy deposits coming from the high energy interaction (the signal) and from the underlying event where the amplitude for both interactions peaks at time 0 ns (the current bunch crossing).

  Due to the large separation of the bunches in 2010 LHC running, only in-time pile-up is present in 2010 data.

- **Out-of-time pile-up** is caused by a mixing of the considered high energy event with soft interactions coming from previous/next bunch crossings. The out-of-time pile-up happens if there is a small separation between the proton bunches and a relatively long pulse shape (in TileCal the pulse shape is from -75 to 75 ns). The high energy signal in the calorimetric cell (with an amplitude peak at 0 ns) is superimposed with the energy deposit coming from the nearby bunch crossing (with a peak shifted in time).

  The spacing between the bunches was 50 ns during 2011 data taking, allowing also the out-of-time pile-up to affect the signal reconstruction. The signal in the nominal bunch crossing has an amplitude corresponding to time 0 ns, while the out-of-time pile-up events have a peak at ± 50 ns in this case.

3.3.2 Monte Carlo Simulations with Pile-up

The Monte Carlo simulations of events in the environment with pile-up are done by mixing the high energy signal event with an appropriate number of simulated minimum bias events. The pile-up conditions in the Monte Carlo simulations are defined by the separation between the bunch crossings (Δt) and the average number of pile-up collisions per bunch crossing (μ). The number of pile-up collisions per bunch crossing follows the Poisson distribution with the mean value of μ.

The Pythia generator [9] is used to produce the underlying interactions. The energy deposits coming from the high energy signal and from the minimum bias collisions are simulated separately. The simulations in both cases follow the standard digitisation procedure in the Tile Calorimeter up to the derivation of the seven samples after 25 ns as described in Sect. 3.1.3. Next, the samples coming from the minimum bias interactions are added to the appropriate samples generated by the high energy collision. After the application of the pedestal and electronic noise, the merged samples enter the signal reconstruction by means of the Optimal Filtering method which is described in Sect. 3.1.2. After the application of the calibration constants (conversion

3.3 Multiple Interactions: Pile-up

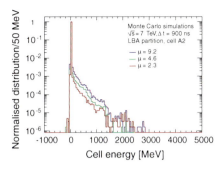 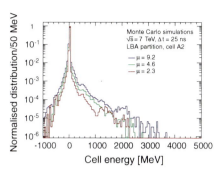

Fig. 3.10 Normalised cell energy distribution for a cell A2 in the LBA partition ($\eta = 0.15$) integrated over ϕ for different pile-up conditions with μ varying from 2.3 to 9.2. The bunch spacing of 900 ns (only in-time pile-up) is used in the *left hand side* plot and of 25 ns (also out-of time pile-up contribution) on the *right hand side*. The electronic noise was switched off in these special simulations

factors from ADC counts to GeV) to the reconstructed amplitude, the cell energy at the EM scale is derived.

Special Monte Carlo simulations have been produced to enable the observation of the pile-up effect only on the cell energy spectrum. The electronic noise is assumed to be zero in these special simulations. All figures shown in the following (Sects. 3.3.2 and 3.3.3) have been prepared using these special Monte Carlo samples.

The cell energy distribution in different pile-up conditions for one of the TileCal cells in the first radial layer (cell A2 in the LBA partition corresponding to $\eta = 0.15$) is shown in Fig. 3.10. The pile-up contribution is assumed to be independent on the azimuthal angle, that is why the ϕ coordinate is not specified. The energy spectra coming from pile-up collisions are highly asymmetric with long tails in the positive energy part. As expected, the pile-up contribution increases with μ. The pile-up configuration with the out-of-time mixing (*right hand side figure*) leads to a wider distribution with a significant tail also in the negative cell energy compared to the in-time pile-up only case (*left hand side figure*).

3.3.3 Pile-up Constants

Since the pile-up influences the cell energy reconstruction, it might affect also the formation of the clusters which are important e.g. for the jet reconstruction. The energy deposits coming from the minimum bias collisions could lead to a cell energy higher than a 4σ equivalent of the cell electronic noise and a fake cluster seed would be created in this case. This undesired behaviour is avoided by adding the pile-up contribution to the electronic noise to define the cell energy spread in the presence of pile-up, denoted as $\sigma_{el.noise+pileup}$. Then the variable $\sigma_{el.noise+pileup}$ is used to define 2σ and 4σ limits in the clustering algorithm under the considered pile-up conditions. The pile-up contribution is expected to be uncorrelated with the electronic noise and therefore it is added in quadrature to the electronic noise cell by cell

$$\sigma_{\text{el.noise+pileup}} = \sqrt{\sigma_{\text{pileup}}^2 + \sigma_{\text{el.noise}}^2} \qquad (3.7)$$

The dependence of the pile-up contribution on the pseudorapidity has been studied in the Monte Carlo simulations. The pile-up is expected to be symmetric in η and independent on the azimuthal angle. Thus the dependence of the cell energy spread (σ_{pileup}) on $|\eta|$ integrated over ϕ for three radial layers is shown in Fig. 3.11. Simulations with two specific pile-up conditions with fixed $\mu = 2.3$, but different bunch spacing ($\Delta t = 900$ ns and $\Delta t = 25$ ns) are used. The pile-up contribution decreases significantly with the radial distance from the beam axis (from layer A to layer D) as expected for soft underlying interactions coming from the primary vertex. The out-of-time pile-up leads to a broader energy distribution, i.e. larger σ_{pileup} values, as already shown for a typical cell in Fig. 3.10.

Variations of the variable σ_{pileup} for the fixed bunch spacing ($\Delta t = 25$ ns), but for different average number of pile-up collisions in the layer A are shown in Fig. 3.12. The dependence is plotted for low luminosity and high luminosity scenarios. Whereas the low luminosity case is simulated at $\sqrt{s} = 7$ TeV with μ between 2.3 and 9.2 (*left hand side*), the high luminosity pile-up is simulated at $\sqrt{s} = 14$ TeV with μ between 23 and 46 (*right hand side*).[4] The pile-up contribution grows rapidly with the increasing number of average minimum bias collisions per bunch crossing.

The $|\eta|$-dependent pile-up constants are derived for both data and Monte Carlo simulations. The constants are specific for the considered pile-up conditions defined by Δt and μ. In order to avoid changing the values in the database for each pile-up configuration, an approximate relation reducing the dependence of σ_{pileup} on the average number of minimum bias collisions is used. It is assumed that the pile-up

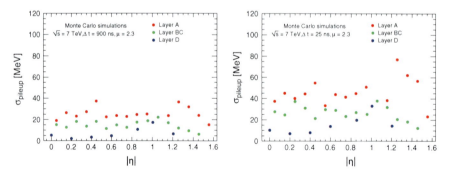

Fig. 3.11 Spread of the cell energy (σ_{pileup}) as a function of $|\eta|$ for collisions in two different pile-up scenarios. Both simulations have been performed at the centre of mass energy of 7 TeV with $\mu = 2.3$. The bunch spacing of 900 ns (only in-time pile-up) is used in the *left hand side* plot and of 25 ns (also out-of time pile-up contribution) in the *right hand side* plot

---

[4] The simulations at the centre of mass energy 14 TeV have been produced for dedicated upgrade studies.

3.3 Multiple Interactions: Pile-up

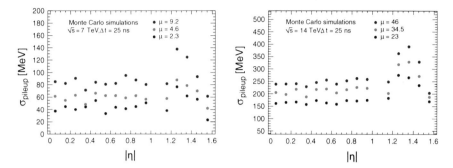

Fig. 3.12 Pile-up contribution σ_{pileup} in the layer A as a function of $|\eta|$ for low luminosity and high luminosity case with bunch spacing of 25 ns. The low luminosity case is simulated at $\sqrt{s} = 7$ TeV with μ between 2.3 and 9.2 (*left hand side*). The high luminosity pile-up is simulated at $\sqrt{s} = 14$ TeV with μ between 23 and 46 (*right hand side*)

contribution scales with μ for the fixed bunch spacing like

$$\sigma_{pileup} = \sigma_{pileup}^{norm} \times \sqrt{\mu} \tag{3.8}$$

where σ_{pileup}^{norm} denotes so-called pile-up constant. The scaling with $\sqrt{\mu}$ is not exact, but it leads to a good agreement in the case of close enough pile-up conditions and for high number of minimum bias interactions.

The pile-up constants σ_{pileup}^{norm} are stored in the database and are used together with the electronic noise constants to derive 2σ and 4σ limits in the clustering algorithm as mentioned above.

References

1. The ATLAS Collaboration, G. Aad et al., Readiness of the ATLAS Tile Calorimeter for LHC collisions. EPJC **70**, 1193 (2010)
2. J. Poveda et al., ATLAS TileCal read-out driver system production and initial performance results. IEEE Trans. Nucl. Sci. **54**, 2629–2636 (2007)
3. The GEANT4 Collaboration, S. Agostinelli et al., GEANT4: a simulation toolkit. Nucl. Instrum. Meth. **A506**, 250 (2003)
4. The ATLAS Collaboration, G. Adranga et al., Testbeam studies of production modules of the ATLAS Tile calorimeter. Nucl. Instrum. Methods Phys. Res. **606**, 362–394 (2009)
5. W. Lampl et al., Calorimeter clustering algorithms: description and performance, ATL-LARG-PUB-2008-002, (2008)
6. R. Brun, F. Rademakers, ROOT - An object oriented data analysis framework. Nucl. Instrum. Methods Phys. Res. **389**, 81–86 (1997)
7. The ATLAS Collaboration, G. Aad et al., Luminosity Determination in pp Collisions at $\sqrt{s} = 7$ TeV using the ATLAS Detector at the LHC. Eur. Phys. J. **C71**, 1630 (2011)
8. The ATLAS Collaboration, Luminosity Public Results (2011), https://twiki.cern.ch/twiki/bin/view/AtlasPublic/LuminosityPublicResults
9. T. Sjostrand, S. Mrenna, P. Skands, PYTHIA 6.4 physics and manual. JHEP **05**, 026 (2006)

Chapter 4
Electron Efficiency Measurement

Electrons play an important role in many physics analyses performed with the ATLAS detector, both in Standard Model measurements and in new physics searches. The electron might also appear in the $Z \rightarrow \tau\tau$ decay, which is discussed in Chap. 5, in the case when one of the τ leptons decays into an electron and two neutrinos. The measurements described in this chapter are performed following the electron selection specific for the $Z \rightarrow \tau\tau$ analysis and the results are used for the cross section measurement in this channel.

The electron reconstruction and identification criteria are summarised in the first part of this chapter (Sect. 4.1). Next, the methodology of the electron efficiency measurements by the so-called tag and probe method is introduced (Sect. 4.2). Then efficiency measurements of the identification cuts (Sect. 4.3), the trigger requirement (Sect. 4.4) and the isolation criteria (Sect. 4.5) in both data and Monte Carlo simulations are described.

4.1 Electron Reconstruction and Identification

The electron reconstruction, identification and its performance in collisions' data with the ATLAS detector are described in details in Ref. [1]. A brief summary is given below.

4.1.1 Reconstruction

The electron reconstruction in the central region ($|\eta| < 2.47$) starts from clusters in the electromagnetic (EM) calorimeter which are associated to tracks coming from the charged particles in the inner detector. Since the forward electrons are not used in

J. Nováková, *Standard Model Measurements with the ATLAS Detector*,
Springer Theses, DOI: 10.1007/978-3-319-00810-3_4,
© Springer International Publishing Switzerland 2014

the $Z \to \tau\tau$ analysis, only the "central electrons" (i.e. electrons within $|\eta| < 2.47$) are discussed in this chapter.

The seed clusters with transverse energies above 2.5 GeV are found by a sliding window algorithm which searches among calorimeter towers with the size of 3×5 cells in the $\eta \times \phi$ plane (one cell corresponds to $\Delta\eta \times \Delta\phi = 0.025 \times 0.025$ in the middle layer of the EM calorimeter). Duplicate clusters are removed based on the energy comparison of the nearby seed clusters. The reconstructed tracks in the inner detector are matched to the clusters in the next step. In the case that multiple tracks are matched to the same cluster, tracks with silicon hits (hits in the Pixel/SCT subdetectors of the inner tracker) have higher priority than tracks with TRT hits only (i.e. tracks without any silicon hits) which are more likely to belong to electrons originating from photon conversions. The track with the smallest distance $\Delta R = \sqrt{\Delta\eta^2 + \Delta\phi^2}$ between its impact point in the EM calorimeter and the seed cluster coordinates is selected. Finally, the electron cluster is rebuilt using towers of 3×7, resp. 5×5 cells in the barrel, resp. endcaps.

The energy of the final cluster is corrected to account for energy deposited outside the cluster region. Energy deposits in the material in front of the EM calorimeter, energy deposits outside the given cluster size inside the EM calorimeter (lateral leakage) and energy deposits beyond the EM calorimeter (longitudinal leakage) are estimated and used for the calculation of the electron transverse energy. On the contrary, the directions η and ϕ of the electron are not taken from the cluster variables but the track parameters at the vertex are considered.

The reconstructed electrons (the reconstructed cluster with the associated track) are a mixture of prompt electrons and electrons coming from a photon conversion, since both can be characterised by the associated track. The identification criteria, discussed below, are applied on the reconstructed electrons to separate the prompt electrons and enhance the purity of the electron sample.

4.1.2 Identification

The electron identification in ATLAS in 2011 is performed with a cut-based approach. The calorimeter, tracking and combined variables are used for the selection. The identification criteria are optimised to provide a good separation between the signal and background electrons as well as jets faking electrons. The background electrons come mainly from the photon conversion or Dalitz decays of π^0 ($\pi^0 \to e^+e^-\gamma$, BR = 1.12 %). The jets might contain real electrons from the B-hadron decays. These electrons are primarily not isolated from other particles inside the jet contrary to the "signal electrons" that are usually isolated.

Three sets of cuts with increasing background rejection power and decreasing signal efficiency are defined: *loose*, *medium* and *tight* with an expected jet rejection power of about 500, 5,000 and 50,000 based on the Monte Carlo simulations. A brief summary of the identification criteria is given below, more details can be found in

Ref.[1]. Only calorimetric information (e.g. lateral width of the shower in the middle layer of the calorimeter, hadronic leakage-ratio of E_T in the hadronic calorimeter and in the EM cluster) is used for the loose selection. More shower width variables (e.g. total shower width in the first layer of the EM calorimeter), track quality requirements (e.g. number of hits in the Pixel and SCT detectors, transverse impact parameter) and the matching of the track and the cluster (requirement on $\Delta\eta$ between the track and the cluster) are added to form the medium selection. The tight selection adds E/p ratio and particle identification using information from TRT detector (e.g. ratio of high threshold TRT hits to the total number of hits in TRT). The tight selection reduces the number of electrons coming from conversions using e.g. a cut on the minimum number of hits in the B-layer (the innermost layer of the Pixel detector barrel region). The selection cuts are optimised in several η and E_T bins in the range of the electron transverse energy from 5 GeV to approximately 100 GeV.

Electrons passing tight identification criteria with $E_T > 17$ GeV in the central region of the detector ($|\eta| < 2.47$) are used in the $Z \rightarrow \tau\tau$ analysis. Hence especially the electrons passing tight identification cuts in the relevant energy region are discussed in more details below.

4.2 Methodology for the Electron Efficiency Measurements

The electron efficiency measurements represent an important part for the cross section measurements because the measured electron spectrum has to be corrected for efficiencies related to the electron selection. According to Ref.[1], the correction factor is defined as a product of different efficiency terms and it can be written in the case of the single electron in the final state as follows

$$C = \epsilon_{event} \cdot \alpha_{reco} \cdot \epsilon_{ID} \cdot \epsilon_{trig} \cdot \epsilon_{isol} \tag{4.1}$$

where ϵ_{event} stands for the efficiency of the event selection cuts, α_{reco} denotes the reconstruction efficiency to find an electromagnetic cluster and to match it to a reconstructed track in the required kinematic range, ϵ_{ID} means identification efficiency with respect to all reconstructed electron candidates, ϵ_{trig} and ϵ_{isol} represent trigger and isolation cuts efficiency with respect to all reconstructed electrons passing considered identification criteria. The measurement of ϵ_{ID} is described in details in Sect. 4.3, ϵ_{trig} in Sect. 4.4 and ϵ_{isol} in Sect. 4.5. The measurement of α_{reco} is not discussed in this chapter, but details can be found in Ref.[1].

The electron efficiency measurements are performed by means of a so-called tag and probe (T&P) method. The T&P method is based on finding a clean sample of real electrons (called *probe* electrons) using a specific selection cuts (called *tag* requirements) applied on another object in the event. A well-identified electron is used as the tag in the $Z \rightarrow ee$ and $J/\psi \rightarrow ee$ events and a high missing transverse energy in the $W \rightarrow e\nu$ T&P method. The efficiency of any electron selection cut (e.g. identification cuts, isolation requirements, electron trigger) can be studied on the sample of the probe electrons.

After the event selection is done and the probe electrons are found, the probe sample might suffer from a background contamination coming primarily from the jets reconstructed as electron candidates. This is the case of the electron identification efficiency measurement where the probe electrons are all reconstructed electrons (the reconstructed cluster with the associated track without any identification requirements stands for the reconstructed electron). Therefore a dedicated background subtraction has to be performed. After the background subtraction is applied, the number of all probes (N_{probe}) and the number of probes passing the considered selection cut (N_{pass}) are derived. The selection efficiency ϵ is calculated as a fraction of probe electrons passing the required criteria ($\epsilon = N_{pass}/N_{probe}$).

The measurements are performed in both data and Monte Carlo simulated samples. Differences between the efficiencies in data and Monte Carlo simulations have been found. The differences come mainly from the description of the electron shower shapes in the Monte Carlo simulations. Correction factors defined as the ratio of efficiency measured in data and in Monte Carlo simulations and ($\epsilon_{data}/\epsilon_{MC}$) are derived in bins of electron η and E_T. These factors, so-called scale factors, are applied on the Monte Carlo efficiency predictions in order to come to an agreement with real data measurements. More details about the usage of the scale factors in the specific physics analysis ($Z \rightarrow \tau\tau$ cross section measurement) can be found in Sects. 5.3.2 and 5.6.1.

4.3 Electron Identification Efficiency with W Tag and Probe Method

The measurement of the electron identification efficiencies and scale factors with W T&P method is described in this section. The identification efficiency is defined as a fraction of reconstructed electrons passing the considered identification criteria (loose, medium or tight). An additional track quality cut has to be required for the reconstructed electrons to suppress the beam-halo background. More precisely, the identification efficiency is defined as $\epsilon_{ID} = N_{pass}/N_{probe}$ where

- N_{probe} is the number of reconstructed electrons, i.e. the reconstructed cluster with the associated track without any identification requirements, with the additional track quality cut
- N_{pass} is the number of reconstructed electrons with the additional track quality cut passing the considered identification cuts (tight criteria in case of $Z \rightarrow \tau\tau$ analysis).

Since no identification cuts are applied to probe electrons, the contribution of background electrons might be high for the probe electrons. Therefore the background subtraction is the key point in the identification efficiency measurement.

4.3 Electron Identification Efficiency with W Tag and Probe Method

4.3.1 Data and Monte Carlo Samples

Data samples collected during the first part of 2011 data taking period are used in the W T&P analysis. Data collected at $\sqrt{s} = 7$ TeV with stable beam conditions in the subset of luminosity blocks with no serious problems in the various subsystems are considered. The analysed data sample corresponds to the integrated luminosity of approximately 2.1 fb^{-1}.

The $W \rightarrow e\nu$ signal Monte Carlo samples are generated with the Pythia generator [2] and processed through the full detector simulation based on the Geant4 simulation programme [3]. The average number of minimum bias collisions per bunch crossing is approximately six during the considered data taking period [4]. The pile-up is also simulated in the Monte Carlo samples, but the average number of interactions per bunch crossing does not correspond to the values measured in data. Therefore a special re-weighting procedure is applied to the Monte Carlo simulated samples to agree with the pile-up conditions measured in data.

4.3.2 Event Selection

The selection of $W \rightarrow e\nu$ events is described in this section. The standard event cleaning procedure used in ATLAS experiment is introduced. The triggers used for selecting the W events are discussed. Then analysis-specific cuts suppressing the most important backgrounds (mainly multijet events) are presented.

Event Cleaning

The event cleaning and the selection of good collision events is performed as the first step of the event selection. The collision events are selected by requiring a primary vertex with at least three associated tracks. Furthermore, an additional check on the quality of jets in the event is performed. The event is rejected if a problematic jet is found as it might affect the measurement of the missing transverse energy (E_T^{miss}) which is crucial for the W T&P analysis.

Finally, the quality check on the reconstructed electrons is also done. Due to the fact that the electromagnetic calorimeter experienced hardware problems during the 2011 data taking period leading to an acceptance hole in the calorimeter, the electron is rejected if it is localised in the region of the readout problems. The acceptance hole is not simulated in the Monte Carlo samples used in the analysis and a correction factor accounting for the acceptance lost is applied to the simulated samples.

Trigger Requirement

Since the single electron trigger might bias the selection of the probe electrons, the missing transverse energy triggers have to be used in the $W \to e\nu$ T&P analysis. However, the development and usage of the E_T^{miss} triggers is challenging in the high luminosity environment at the LHC. A set of various missing transverse energy triggers available during 2011 data taking, selecting always unprescaled triggers with lowest possible threshold, is used.

The triggers based on the E_T^{miss} significance, where the E_T^{miss} resolution is parametrised and a threshold for the ratio of E_T^{miss} over its significance is set, are considered in the first periods of 2011 data taking. Later in 2011, a loose requirement on an electron candidate track and a minimal distance of the missing transverse energy from a nearby jet object has to be added to the E_T^{miss} significance requirement to cope with increasing luminosity conditions. The list of triggers used in data is summarised in Table 4.1.

Due to the rapid changes in the trigger configuration during 2011 data taking period, the trigger setup in Monte Carlo simulated samples is not identical to the configuration in data. Moreover, some of the triggers used for data collected during 2011 are missing in the simulated samples. The trigger collection used for the Monte Carlo simulated events is composed from the available triggers in a way to correspond to the appropriate fractions of events passing the given trigger in data.

The triggers used in the W T&P studies are defined especially for allowing this method to be used with 2011 data and therefore it is not expected to bias the W T&P selection. Nevertheless, the possible bias of the efficiency measurement on the trigger type has been studied and it is found to be negligible. More details are given in Sect. 4.3.5.

Table 4.1 List of triggers at the event filter (EF) level used for the W T&P analysis with their fractions measured in data after the whole event selection

Trigger	Number of probes	Fraction (%)
EF_xs60_noMu_L1EM10XS45	3.1×10^5	19.3
EF_xs75_noMu_L1EM10XS50	9.4×10^3	0.6
EF_g20_etcut_xe30_noMu	1.5×10^4	0.9
EF_e13_etcut_xs60_noMu	7.8×10^5	47.9
EF_e13_etcut_xs60_noMu_dphi2j10xs07	5.1×10^5	31.2

The fraction of events with the given trigger is calculated assuming the priority of the triggers given by the trigger threshold, corresponding to the position in the table. Briefly, the meaning of the labels is the following: xs60, resp. xs75 stands for the E_T^{miss} significance cut, noMu means that no correction for the possible presence of muons is applied, g20 looks for a photon with $E_T > 20$ GeV at EF, xe30 means missing transverse energy cut, e13_etcut requires a calorimeter cluster with $E_T > 13$ GeV with a good-quality track pointing to it, dphi2j10xs07 represents a cut on the minimal azimuthal distance of the E_T^{miss} vector and the jet at 0.7 similar to the cut $\Delta\phi$ used also in the offline selection described in the text

4.3 Electron Identification Efficiency with W Tag and Probe Method

Event Selection Cuts

The event selection is optimised to find a clean sample of $W \to e\nu$ events without applying any identification cut on the reconstructed electron. Due to the large background coming from jets reconstructed as electron candidates at low electron energies, the efficiency measurements with W T&P method are performed for reconstructed electrons with $E_T > 15$ GeV. As already mentioned, only "central electrons" with $|\eta| < 2.47$ are considered in the analysis.

First, a so-called Z boson veto is applied in order to reduce the contamination from the Z+jets background. The event is rejected if two or more electrons passing medium identification criteria with $E_T > 15$ GeV are found.

Then a cut on the missing transverse energy is performed. The reconstruction of the missing transverse energy uses the energy deposits in the calorimeter and the reconstructed muon tracks.[1] The missing transverse energy has to satisfy $E_T^{\text{miss}} > 25$ GeV.

Next, a transverse mass of the W candidate (m_T) is a good discriminant variable between $W \to e\nu$ events and the multijet background

$$m_T = \sqrt{2E_T(e) \cdot E_T^{\text{miss}} \cdot [1 - \cos \Delta\phi(e, E_T^{\text{miss}})]} \,. \tag{4.2}$$

The multijet background events can be characterised by low values of m_T contrary to the signal events. Events passing $m_T > 40$ GeV are selected.

Furthermore, a special cut to reduce the multijet background is applied. The fake electrons coming from multijet events can be reduced to a large extent by requiring the E_T^{miss} vector to be isolated from the jets in the event [1]. The difference $\Delta\phi$ between the azimuthal angles of the missing transverse energy vector and any reconstructed jet with $E_T > 10$ GeV is required to be larger than 2.5. The strict cut is chosen especially to reduce significantly the multijet background in the low E_T region ($E_T < 25$ GeV) even for the price of the non-negligible signal loss. Another motivation for this cut-value is the trigger evolution. At the end of 2011 data taking period the missing energy trigger used for W T&P contains the $\Delta\phi$ cut at 2.0 and a higher cut value is used in the offline selection.

Finally, due to the presence of the beam-halo background muons producing high-energy bremsstrahlung clusters in the EM calorimeter, track quality requirements (at least one pixel hit and a total of at least seven pixel/SCT hits on the reconstructed track) are applied to all reconstructed electrons [1].

The efficiency of the individual event selection cuts for data corresponding to approximately 2.1 fb^{-1} is shown in Table 4.2. Only events passing the whole cutflow chain are considered in the analysis. The reconstructed electrons in the given

---

[1] The missing transverse energy is calculated as a vector sum $\vec{E}_T^{\text{miss}} = \vec{E}_T^{\text{miss}}(\text{calo}) + \vec{E}_T^{\text{miss}}(\text{muon}) - \vec{E}_T^{\text{miss}}(\text{energy loss})$, where $\vec{E}_T^{\text{miss}}(\text{calo})$ is evaluated from the energy deposits in the calorimeter cells inside topological clusters, $\vec{E}_T^{\text{miss}}(\text{muon})$ is the sum of the muon momenta and $\vec{E}_T^{\text{miss}}(\text{energy loss})$ is a correction term accounting for the muons' energy lost in the calorimeters.

Table 4.2 Number of data corresponding to approximately 2.1 fb^{-1} during the event selection chain

	Number of events	Relative acceptance (%)
Trigger and event cleaning	2.020×10^7	100.0
Electron kinematic cuts	1.733×10^7	85.8
Z boson veto	1.732×10^7	100.0
$E_T^{miss} > 25$ GeV	1.397×10^7	80.7
$m_T > 40$ GeV	1.363×10^7	97.6
$\Delta\phi(E_T^{miss}, \text{jet}) > 2.5$	1.808×10^6	13.3
Track quality	1.619×10^6	89.5
Number of probes	1.619×10^6	100.0
Number of loose electrons	1.402×10^6	86.6
Number of medium electrons	1.342×10^6	82.9
Number of tight electrons	1.106×10^6	68.4

The relative acceptance is given with respect to the previous cut. The numbers under relative acceptance for electrons passing loose, medium and tight identification criteria are raw efficiencies with no background subtraction applied. Therefore they do not represent any measurement of the efficiency, but it is for information only

kinematic region passing the track quality cuts form the probe electrons and the efficiency of the electron identification cuts can be measured on this sample.

Various kinematic variables for electrons passing tight identification cuts, where the background contamination in data is negligible, in both data and Monte Carlo simulations are shown in Fig. 4.1. The whole selection chain is applied and no background subtraction is done at this step. The overall agreement between data and Monte Carlo is good and the residual differences in the electron E_T spectrum, m_T and E_T^{miss} distributions can be explained by different triggers used in data and in Monte Carlo simulations.

4.3.3 Background Subtraction

A clean sample of real electrons at the probe level is very important for a precise measurement of the electron identification efficiency. The largest contamination of the jets being mismeasured as an electron candidate is expected in the low energy region ($E_T < 25$ GeV) where a careful treatment of the background is especially required.

The observable suitable for discriminating the isolated signal electrons from the jets faking electrons is required to be defined without using any of the electron identification criteria. The calorimeter isolation of the electron was chosen as the discriminating variable [1]. Even though being slightly correlated to some of the electron identification variables (e.g. hadronic leakage R_{had}-ratio of E_T in the hadronic calorimeter and in the EM cluster), it has been found to be the best choice. The calorimeter isolation is defined as a sum of the transverse energies of all cells in the

4.3 Electron Identification Efficiency with W Tag and Probe Method

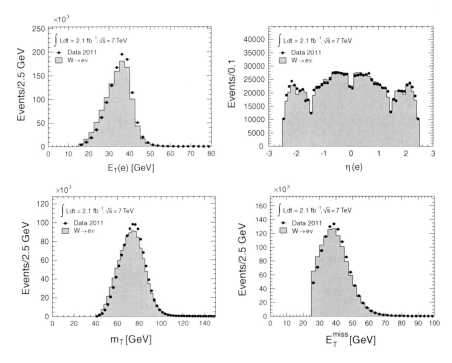

Fig. 4.1 Kinematic distributions for the tight electrons selected by the W T&P method in data and in Monte Carlo simulations. No background subtraction is performed at this stage. The Monte Carlo simulated samples are normalised to data in all figures

EM and hadronic calorimeter within the given cone size ($\Delta R = 0.3$ or 0.4) in the $\eta - \phi$ space, the energy deposits in cells associated to the electron cluster itself are subtracted from the sum. Finally, the sum is normalised to the transverse energy of the electron to minimise the E_T dependence of the isolation variable. The calorimeter isolation with the cone size $\Delta R = 0.3$, resp. 0.4 is denoted as $E_T\text{Cone}30/E_T$, resp. $E_T\text{Cone}40/E_T$ in the text.

The real electrons from $W \to e\nu$ decay are preferably isolated and the calorimeter isolation is expected to peak at values close to zero. The width of the distribution is given by the contributions of the electronic noise, pile-up and shower leakage. On the contrary, fake electrons from multijet background would form a much broader distribution with values even larger than one. The spectrum of the $E_T\text{Cone}40/E_T$ variable is shown in Fig. 4.2 for both data and Monte Carlo simulations. The long tail in the region with large isolation values for probe electrons in data comes from the residual multijet background. The tail vanishes when the tight identification criteria are applied to the reconstructed electrons. A good agreement between the probe electrons and electrons passing tight identification criteria can be observed in the low isolation region where the background contribution is small. On the other hand, the agreement between data and Monte Carlo shows some discrepancies. The Monte

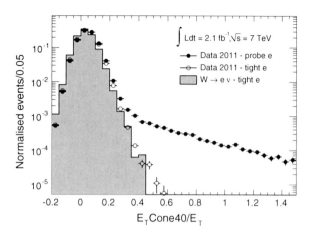

Fig. 4.2 Spectrum of the isolation variable $E_T\text{Cone}40/E_T$ for probe electrons, electrons passing tight identification cuts in data and tight electrons from $W \rightarrow e\nu$ Monte Carlo. All distributions are normalised to unity. Electrons with 35 GeV $< E_T <$ 40 GeV in the whole pseudorapidity region except the crack ($1.37 < |\eta| < 1.52$) are considered. The long tail to large values of $E_T\text{Cone}40/E_T$ for probe electrons is caused by fake electrons coming from the multijet background

Carlo distribution is slightly shifted to lower values of the isolation variable in the region of the peak. However, the isolation variable is rather a handle to discriminate signal electrons from fake electrons coming from the multijet background in data and no direct comparison between data and Monte Carlo is performed in the analysis. Therefore the absolute agreement in the isolation distribution between data and Monte Carlo is not crucial for the efficiency measurements.

The background estimation is done using data driven background templates. The background templates are constructed separately for probe electrons and for electrons passing identification criteria. Both background templates are built by selecting such probe electrons that are likely to be fake electrons coming from the multijet background. The set of fake electrons is built by requiring the reconstructed electrons to fail certain identification cuts, i.e. the total shower width in the first layer of the EM calorimeter, ratio of high threshold TRT hits to the total number of hits in TRT and ratio in ϕ of cell energies in the middle layer of the EM calorimeter. Generally, the set of variables used for building the background template is chosen by two main requirements. First, a good agreement in the shape of the electron calorimeter isolation outside the signal region between the background template and the data is required. Second, a reasonable statistics of the background template in all η and E_T bins is necessary. A robust background template is built by requiring the probe electrons to fail two out of three identification variables mentioned above. Two different background templates are defined in this way and they are used in the calculation of the systematics as described in Sect. 4.3.5. The template for electrons passing the identification cuts (the numerator in the efficiency calculation) is built in the same way and in addition the electrons have to fulfil the R_{had} cut. The hadronic leakage

4.3 Electron Identification Efficiency with W Tag and Probe Method   41

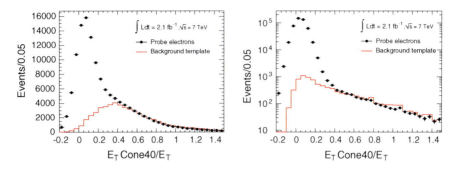

Fig. 4.3 Calorimeter isolation (E_TCone40/E_T) for probe electrons in two E_T bins: 20 – 25 GeV on the *left hand side* and 35 – 40 GeV on the *right hand side* (note the logarithmic scale). Electrons in the whole considered pseudorapidity region excluding the crack region ($1.37 < |\eta| < 1.52$) are used. The background templates, defined by inverting the cuts on the total shower width in the first layer of the EM calorimeter and the ratio of high threshold TRT hits to the total number of hits in TRT, are also shown in the figures

is highly correlated with the electron isolation and this cut is used in the medium and tight selection. Adding this cut is necessary to obtain a good agreement in the isolation shapes for electrons passing the identification criteria.

The last step to derive the number of signal events is to normalise correctly the background template and subtract the estimated background from data. A signal region is defined by a certain isolation threshold, e.g. 0.4. The background templates are expected to be signal free outside the signal region and are normalised to the data in this region. The background templates are used to subtract the residual background in the signal region after the normalisation is applied.

An example of the background templates for probe electrons in two electron E_T bins is shown in Fig. 4.3. The significant level of the background in the low E_T region ($20 < E_T < 25$ GeV) can be seen. On the contrary, the bin with electrons within $35 < E_T < 40$ GeV has a very low background level already for probe electrons. The background template agrees well with data outside the signal region as required.

The number of signal and background electron probes in data as well as signal over background ratios (S/B) in different E_T bins can be found in Table 4.3. The largest contribution of the background is in the lowest E_T bins ($E_T < 25$ GeV) where the S/B ratio is smallest. The signal region becomes very clean with increasing electron energy. The highest statistics is available in the region $35 < E_T < 40$ GeV.

4.3.4 Efficiency and Scale Factors Measurement

The number of probes and electrons passing the identification criteria are extracted after the background subtraction procedure is performed. The efficiency of the iden-

Table 4.3 Number of signal and background probes and signal over background ratios (S/B) in the signal region (E_TCone40$/E_T < 0.4$) in different E_T bins

E_T (GeV)	Signal	Background	S/B
15–20	16132 ± 260	11174 ± 201	1.44 ± 0.03
20–25	71439 ± 467	23339 ± 351	3.06 ± 0.05
25–30	198573 ± 634	21807 ± 427	9.1 ± 0.2
30–35	369511 ± 714	12478 ± 358	29.6 ± 0.9
35–40	463429 ± 722	5582 ± 227	83.0 ± 3.4
40–45	197036 ± 462	2162 ± 118	91.1 ± 5.0
45–50	34740 ± 198	825 ± 61	42.1 ± 3.1

Electrons over the whole pseudorapidity region excluding the crack region ($1.37 < |\eta| < 1.52$) are summed up. The errors are statistical only. The data sample corresponds to 2.1fb^{-1}

tification cuts in both data and signal Monte Carlo samples[2] are measured in bins of the electron pseudorapidity and transverse energy. The η-dependent efficiencies and scale factors are derived using electrons with 20 GeV $< E_T < 50$ GeV, whereas the E_T-dependent values are integrated over the whole pseudorapidity region excluding the crack region ($1.37 < |\eta| < 1.52$). The measured efficiencies and scale factors ($\epsilon_{\text{data}}/\epsilon_{\text{MC}}$) for electrons passing tight identification criteria are shown in Fig. 4.4.

The efficiencies for tight electrons are not uniform as a function of pseudorapidity. It is caused by the fact that the tight identification requires the tracking information and is therefore sensitive to interactions of the electron with the inner detector material.

The identification efficiencies show differences between data and Monte Carlo simulations which can be quantified by the scale factors ($\epsilon_{\text{data}}/\epsilon_{\text{MC}}$). The largest deviation between data and Monte Carlo predictions is observed in the crack region ($1.37 < |\eta| < 1.52$) where a precise description of the material is very difficult. The differences in other η and E_T bins can be explained by imperfect simulations of some of the electron identification variables. Especially the distributions of the variables using the calorimeter shower shapes or high threshold TRT hits do not agree perfectly in data and in Monte Carlo simulations as discussed in Ref. [1]. The E_T dependence of the scale factors is expected to be related to the lateral shower shapes which might differ more from the Monte Carlo expectations in the low E_T range.

---

[2] So-called loose truth matching of the electron is required in the Monte Carlo samples. The loose truth matching includes these cases: First, the electron track in the inner detector is directly matched to the primary electron. Second, electrons are indirectly matched to the true primary electron, meaning that hits in the inner detector correspond to the electron track generated by bremsstrahlung photons or final state radiation photons from the hard process itself. On the contrary, the tracks belonging to charged hadrons or photons conversions from π^0 decays in hadronic jets are omitted by the loose truth matching. This is the desired behaviour since events with a jet being mismeasured as an electron contribute to the so-called the multijet background and are removed by the background subtraction procedure.

4.3 Electron Identification Efficiency with *W* Tag and Probe Method   43

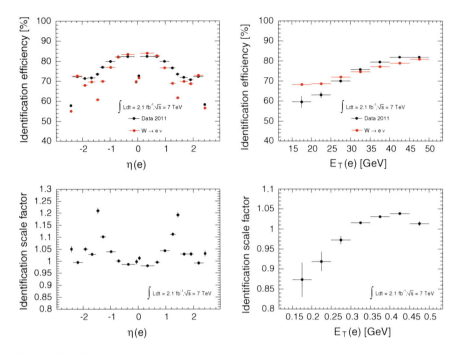

Fig. 4.4 Identification efficiencies (*upper row*) and scale factors (*lower row*) for electrons passing tight identification cuts. The η-dependent efficiencies and scale factors (figures on the *left hand side*) are derived using electrons with 20 GeV $< E_T <$ 50 GeV, whereas the E_T-dependent values (figures on the *right hand side*) are integrated over the whole pseudorapidity region excluding the crack region (1.37 $< |\eta| <$ 1.52). The error bars represent the total uncertainties (statistical and systematic errors summed in quadrature)

4.3.5 Systematic Uncertainties

The main source of the systematic uncertainty on the measured scale factors is coming from the background subtraction which is the crucial part of the identification efficiency measurement. Several variations on the background subtraction are performed to derive the systematic uncertainty associated with it:

- Two different cone sizes ($\Delta R = 0.3, 0.4$) of the calorimeter isolation are used to build the discriminant ($E_T \text{Cone}30/E_T$, $E_T \text{Cone}40/E_T$).
- The threshold defining the signal and background regions in the isolation variable is varied from 0.3 to 0.5.
- Two different background templates are defined by inverting slightly different identification variables.

All these parameters are varied independently resulting in 20 variations in total. Based on the high number of possible variations, none of the configuration is found to be the preferred method and all variations are treated as being equivalent. The

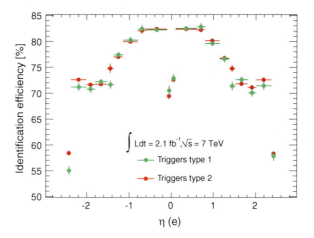

Fig. 4.5 Identification efficiency for tight identification criteria derived with different sets of triggers used in the W T&P analysis. Trigger type 1 stands for the triggers without any electron cluster requirement, type 2 means triggers with a requirement on the calorimeter cluster with a good-quality track. Electrons in the range 20 GeV < E_T < 50 GeV have been used. The error bars represent the total uncertainties (statistical and systematic errors summed in quadrature)

nominal value and the statistical uncertainty in each bin are given by the mean value from all variations in the considered bin. The systematic uncertainty is estimated as the standard deviation of all 20 efficiency measurements.

Moreover, the effect of different triggers used for the efficiency measurements has been studied. The measured efficiencies in data derived with different sets of triggers (requiring a calorimeter cluster with a good-quality track at the EF level or not) are shown in Fig. 4.5. The largest difference is found in the crack region (1.37 < $|\eta|$ < 1.52), but this region is usually not used in the physics analysis. The difference in the η-bin $-2.47 < \eta < -2.37$ is assumed to be a statistical effect due to a good agreement seen in the η-symmetric bin $2.37 < \eta < 2.47$. Otherwise, the measurements with different triggers are in a good agreement and no significant bias from the triggers used to select W T&P events is observed. Therefore no additional systematics related to the set of triggers is assumed.

Pile-up might be another source of systematics in this study. The data used in the analysis were collected during the first part of year 2011 where the average number of pile-up collisions was approximately 6. The identification efficiency is found to be decreasing with the number of pile-up vertices in both data and Monte Carlo simulations. The scale factors are found to be stable against the pile-up in the analysed dataset and no additional systematics related to pile-up is assigned.

The measured efficiencies for tight electrons in data and in Monte Carlo simulations as well as the scale factors together with their statistical and systematic uncertainties in bins of E_T are listed in Table 4.4. The precision measurements in the low E_T region (E_T < 30 GeV) are limited by the systematic uncertainty which is significantly larger than the statistical uncertainty in this region. In particular, in

4.3 Electron Identification Efficiency with W Tag and Probe Method

Table 4.4 Identification efficiencies and scale factors for tight electrons with their statistical (*first uncertainty*) and systematic (*second uncertainty*) errors in different E_T bins

E_T (GeV)	Data (%)	Monte Carlo (%)	Scale factor
15–20	$59.6 \pm 0.9 \pm 2.8$	68.3 ± 0.4	$0.873 \pm 0.014 \pm 0.041$
20–25	$63.1 \pm 0.4 \pm 1.6$	68.6 ± 0.2	$0.919 \pm 0.006 \pm 0.024$
25–30	$70.0 \pm 0.2 \pm 0.7$	71.9 ± 0.2	$0.972 \pm 0.003 \pm 0.010$
30–35	$75.7 \pm 0.1 \pm 0.2$	74.6 ± 0.1	$1.015 \pm 0.002 \pm 0.002$
35–40	$79.5 \pm 0.1 \pm 0.1$	77.1 ± 0.1	$1.031 \pm 0.002 \pm 0.001$
40–45	$81.9 \pm 0.1 \pm 0.1$	78.9 ± 0.2	$1.038 \pm 0.002 \pm 0.001$
45–50	$81.9 \pm 0.3 \pm 0.2$	80.8 ± 0.4	$1.013 \pm 0.006 \pm 0.003$
20–50	$76.6 \pm 0.1 \pm 0.3$	75.0 ± 0.1	$1.021 \pm 0.001 \pm 0.004$

The uncertainty associated to the Monte Carlo simulations comes from the limited statistics of the samples. The integration over the pseudorapidity region excluding the crack region ($1.37 < |\eta| < 1.52$) has been performed

the lowest E_T bin ($15\ \text{GeV} < E_T < 20\ \text{GeV}$) where the statistical uncertainty of $2\,\%$ and systematics of $5\,\%$ occur. The tight identification scale factors systematic uncertainty is $1\,\%$ or smaller for electrons with $E_T > 25\ \text{GeV}$.

4.3.6 Application of the Identification Scale Factors in Physics Analysis

The identification scale factors are used in most of the physics analyses in the ATLAS experiment, e.g. in the $Z \rightarrow \tau\tau$ cross section measurement as described in more details in Sects. 5.3.2 and 5.6.1, to correct the efficiencies predicted by Monte Carlo simulations. The final scale factors which are used as an event weight in the Monte Carlo simulations are derived in two steps. First, the combination of different T&P methods is done in each η and E_T bin. Second, the η and E_T dependent scale factors are multiplied to provide the final scale factor for the considered electron. These two steps are described in more details below.

The η and E_T dependent identification scale factors are measured not only by means of the W T&P method, but also $Z \rightarrow ee$ and $J/\psi \rightarrow ee$ decays are used to derive the efficiencies and scale factors. These three sets of measurements are independent as various triggers, different event selection and background subtraction methods are used. The W and Z T&P measurements are performed in the same kinematic region ($15\ \text{GeV} < E_T < 50\ \text{GeV}$), whereas the J/ψ channel measures the electrons with lower transverse energy ($7\ \text{GeV} < E_T < 20\ \text{GeV}$). The results obtained with different tag and probe methods with 2011 data samples are compared in Ref. [5] showing a reasonable agreement between the measured efficiencies.

The kinematic range used in $Z \rightarrow \tau\tau$ analysis is $E_T > 17\ \text{GeV}$. The J/ψ measurement suffers from rather large systematics in the region $15\ \text{GeV} < E_T < 20\ \text{GeV}$. Thus this measurement is not used for the combination and only the results from W and Z T&P measurements are combined. The combined scale factors in each η and E_T bin are derived by the summation of the individual scale factors weighted by their total uncertainties (statistical and systematic errors summed in quadrature) using the standard error propagation formula. The combination of the W and Z results can be written as

$$SF_{\text{combined}} = \frac{SF_W/\Delta SF_W^2 + SF_Z/\Delta SF_Z^2}{1/\Delta SF_W^2 + 1/\Delta SF_Z^2}, \tag{4.3}$$

$$\Delta SF_{\text{combined}} = \frac{1}{1/\Delta SF_W^2 + 1/\Delta SF_Z^2}. \tag{4.4}$$

The scale factors derived by W and Z T&P methods agree with the combined scale factors within their total uncertainty justifying a good meaning of the combination.

For data in the first part of 2011, the identification scale factors are given separately in bins of η and E_T as shown in the text. The final identification scale factor taking into account both dependencies is given by the product of the η-dependent and E_T-dependent scale factors. The product has to be corrected by an average scale factor (SF_{ID}(average)—scale factor averaged over both η and E_T) in order to avoid double-counting

$$SF_{ID}(\eta, E_T) = SF_{ID}(\eta) \cdot \frac{SF_{ID}(E_T)}{SF_{ID}(\text{average})}. \tag{4.5}$$

These factors are used in the Monte Carlo simulations where for a selected electron passing medium, resp. tight identification criteria appropriate η and E_T dependent scale factors are looked up and used to build the corresponding event weight.

4.4 Single Electron Trigger Efficiency with W Tag and Probe Method

A combined electron and hadronic τ trigger is used in the $Z \rightarrow \tau\tau$ analysis (Chap. 5). The two parts of the trigger are considered to be uncorrelated and the efficiency of the combined trigger is calculated as a product of the two independent efficiencies. This assumption has been checked and confirmed in the $Z \rightarrow \tau\tau$ analysis.

The electron part of the trigger used for the $Z \rightarrow \tau\tau$ analysis, EF_e15_medium, searches for an electron trigger object with $E_T > 15$ GeV at the event filter (EF) level. The measurements of the electron trigger efficiency and scale factors for this particular trigger are performed with respect to electrons passing tight identification criteria which are used in the $Z \rightarrow \tau\tau$ cross section measurement.

The $W \rightarrow e\nu$ tag and probe method is used for the trigger efficiency measurement. The methodology is very similar to the measurement of the identification

4.4 Single Electron Trigger Efficiency with W Tag and Probe Method

efficiency described above. The missing transverse energy triggers listed in Table 4.1 are designed to allow both the electron identification efficiency as well as single electron trigger efficiency measurements with W T&P. The event selection remains the same as described in Sect. 4.3.2, only the definition of probe electrons and electrons passing the required criteria changes. The probe electrons with $E_T > 17$ GeV are required to pass tight identification cuts (N_{probe}), exactly as selected electrons in the $Z \to \tau\tau$ analysis. Then the probability of the probe electrons to be matched to the EF_e15_medium trigger object within the distance of $\Delta R = 0.15$ (N_{pass}) is studied.

Due to the fact that the probe electrons are required to fulfil tight identification requirements, the background contamination is negligible in this study and no background subtraction is performed. The signal region is simply defined by a cut on the electron calorimeter isolation, as it is done in the case of the identification efficiency measurement. The measurements are performed in bins of electron E_T and η. The pseudorapidity and also E_T binning is coarser than the one used for the identification efficiency. Due to the fact that the region close to the trigger cut is difficult to describe in Monte Carlo simulations, the E_T range between 17 and 20 GeV is studied more carefully. Similarly as in the case of the identification scale factors, the measured scale factors are found to depend on the electron E_T apart from the η dependence. The measurement is assumed to be symmetric in η. Thus the absolute value of η is used to enhance the statistics in the lowest E_T bin from 17 to 20 GeV. The scale factors are derived using a 2-dimensional binning of $E_T \times |\eta|$ as shown in Fig. 4.6.

The systematics of the measured trigger efficiency and scale factors is estimated in the same way as for the identification efficiency measurements: Several variations on the selection of the signal region are performed, namely two isolation variables are used ($E_T\text{Cone}30/E_T$ and $E_T\text{Cone}40/E_T$) and the isolation threshold defining the signal region is varied from 0.3 to 0.5. The mean value of the results coming from all variations is taken as the nominal value and the systematic uncertainty is given

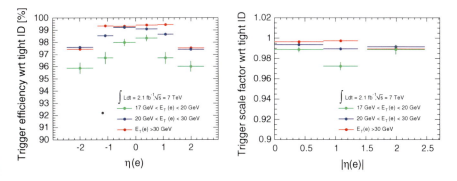

Fig. 4.6 Single electron trigger (EF_e15_medium) efficiency measured in data (*left hand side*) and scale factors (*right hand side*), both with respect to tight identified electrons. The crack region ($1.37 < |\eta| < 1.52$) is removed from the calculation. The error bars represent the total uncertainties (statistical and systematic errors summed in quadrature)

48   4 Electron Efficiency Measurement

by the spread of the individual measurements. The measured efficiencies and scale factors together with their statistical and systematic uncertainties are summarised in Appendix A.

The derived electron trigger scale factors are used in the $Z \rightarrow \tau\tau$ analysis in the same way as the identification scale factors—to correct the Monte Carlo predictions and for the systematics studies. More details are given in Sects. 5.3.2 and 5.6.1.

4.5 Electron Isolation Efficiency with Z Tag and Probe Method

The lepton isolation is used in many physics analyses in order to reduce the multijet background. The isolation of electrons is also used in the $Z \rightarrow \tau\tau$ analysis. The calculation of the electron isolation cuts efficiency specific for the $Z \rightarrow \tau\tau$ analysis is shown in this section.

Several isolation criteria can be defined for electrons. One of them is based on the calorimeter information as already defined in Sect. 4.3.3. The calorimeter isolation in the cone of $\Delta R = 0.4$ is used in the $Z \rightarrow \tau\tau$ analysis. The other isolation variable is using the tracking information. The track isolation is defined in a similar way as the calorimeter isolation: It is calculated as a scalar sum of transverse momentum of the charged particles' tracks in a cone of radius $\Delta R = 0.4$ in $\eta - \phi$ excluding the track belonging to the electron itself. The relative isolation, the isolation variable divided by the electron transverse energy, is considered to reduce the dependence of the isolation variables on the electron momentum. It is denoted as $p_T\mathrm{Cone}40/E_T$ in the following.

The isolation efficiency is analysis-dependent because different isolation requirements are necessary for different analyses. The following isolation criteria applied on electrons passing tight identification criteria are used in the $Z \rightarrow \tau\tau$ analysis (more details in Sect. 5.3.2): $p_T\mathrm{Cone}40/E_T < 0.06$ and $E_T\mathrm{Cone}40/E_T < 0.1$. The measurement of the isolation efficiency by means of the Z T&P method for these specific criteria is shown below.

The tag electron in the $Z \rightarrow ee$ tag and probe method must fulfil $E_T > 20$ GeV cut and be matched to the EF_e20_medium trigger object. Furthermore, the tag electron is required to pass tight identification criteria and an isolation requirement of $E_T\mathrm{Cone}40/E_T < 0.2$ is applied on the tag. The probe electron has to pass tight identification criteria as well and the transverse energy cut is lowered to 17 GeV to correspond to the cut used in the $Z \rightarrow \tau\tau$ analysis. The invariant mass of the tag and probe pair is required to be in the Z boson mass window, namely between 81 and 101 GeV. Due to the very strong requirements on both tag and probe electrons (both passing tight identification cuts), the background contamination is negligible and no background subtraction is necessary in this case.

The isolation efficiency as well as scale factors are found to be dependent on electron pseudorapidity. Moreover, also non-negligible transverse energy dependence of the scale factors is observed as shown in Fig. 4.7. In order to account for both η and

4.5 Electron Isolation Efficiency with Z Tag and Probe Method

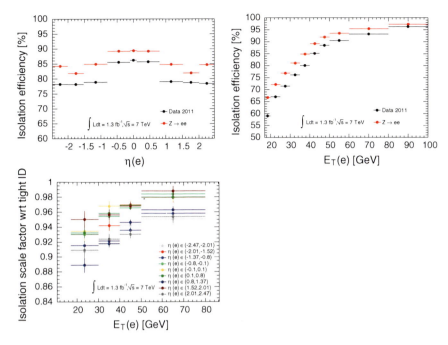

Fig. 4.7 Electron isolation (p_TCone40/E_T < 0.06 and E_TCone40/E_T < 0.1) efficiency and scale factors with respect to electrons passing tight identification criteria. The crack region (1.37 < $|\eta|$ < 1.52) is removed from the calculation. The error bars represent the total uncertainties (statistical and systematic errors summed in quadrature)

E_T dependence, the scale factors are derived using a 2-dimensional binning with 9 × 4 bins in $\eta \times E_T$.

The systematic uncertainty on the derived scale factors comes mainly from the Z T&P method itself. Several variations in the method (e.g. loose requirements on the tag electron—medium identification, no isolation or a change of the invariant mass window) are performed to evaluate the systematics on the scale factors. The individual contributions are assumed to be independent and are evaluated by changing one by one in the nominal selection in contrast with the full variation scan in the previous cases. All shifts from the nominal value are added in quadrature together with the statistical uncertainty to estimate the total uncertainty. The motivation for the simpler treatment of the systematics is the negligible background in this case whereas the background subtraction is the main source of uncertainty in the case of the identification cuts efficiency measurements. The measured efficiencies and scale factors together with their statistical and systematic uncertainties are summarised in Appendix B.

The measured electron isolation scale factors are also used in the $Z \to \tau\tau$ cross section measurement to correct the Monte Carlo predictions and for the evaluation of the systematic uncertainties. More details are given in Sects. 5.3.2 and 5.6.1.

References

1. The ATLAS Collaboration, G. Aad et al., Electron performance measurements with the ATLAS detector using the 2010 LHC proton-proton collision data. EPJC **72**(2012), 1909 (2010)
2. T. Sjostrand, S. Mrenna, P. Skands, PYTHIA 6.4 physics and manual. JHEP **05**, 026 (2006)
3. The GEANT4 Collaboration, S. Agostinelli et al., GEANT4: A simulation toolkit. Nucl. Instrum. Meth. **A506**, 250 (2003)
4. The ATLAS Collaboration, Luminosity public results (2011), https://twiki.cern.ch/twiki/bin/view/AtlasPublic/LuminosityPublicResults
5. The ATLAS Collaboration, G. Aad et al., Identification efficiency measurement for electrons with transverse energy between 7 and 50 GeV, ATL-COM-PHYS-2011-1669 (2011)

Chapter 5
$Z \to \tau\tau$ Cross Section Measurement

The measurement of the $Z \to \tau\tau$ cross section in proton-proton collisions at the centre of mass energy of 7 TeV with the ATLAS experiment is described in this chapter. The analysis has been performed in three different final states determined by the decay mode of the τ lepton. The data sample collected during 2011 corresponding to an integrated luminosity of 1.34–1.55 fb^{-1} depending on the final state is used. The cross section is measured in the $\tau\tau$ invariant mass range $66 < m_{\tau\tau} < 116$ GeV and is documented in Ref. [1].

Since the author of the thesis was involved mainly in the channel with one τ lepton decaying into an electron and neutrinos and the other one hadronically ($Z \to \tau\tau \to e + \text{hadrons} + 3\nu$), so-called electron-hadron channel, this channel is described in more details in this chapter. The other measured final states are mentioned only briefly for completeness. Figures and numbers shown in this chapter are for the electron-hadron channel only.

The organisation of the chapter is following: First, the motivation for the $Z \to \tau\tau$ measurement is given (Sect. 5.1). Next, the signal signature and main background sources are summarised (Sect. 5.2). The object and event selection is described in the next section (Sect. 5.3). Then details about the background estimation are given (Sect. 5.4). Afterwards, the methodology used for the derivation of the cross section is introduced (Sect. 5.5). The estimation of the systematic uncertainties is described next (Sect. 5.6). Finally, the results obtained in the electron-hadron channel are combined with the measurements in the other $Z \to \tau\tau$ final states to derived the total cross section (Sect. 5.7).

5.1 Introduction

The motivation for measuring the $Z \to \tau\tau$ process can be summarised in three items. First, the $Z \to \tau\tau$ process represents a background to some of the Higgs boson searches ($H \to \tau\tau$). On top of that, it is a complementary measurement to

J. Nováková, *Standard Model Measurements with the ATLAS Detector*,
Springer Theses, DOI: 10.1007/978-3-319-00810-3_5,
© Springer International Publishing Switzerland 2014

the Z boson decays into a pair of electrons or muons. Finally, this process plays an important role in the τ performance studies.[1]

Decays with τ leptons in the final state can be a signature of so far unobserved particles. The Standard Model decays of the W and Z bosons with τ leptons ($W \rightarrow \tau\nu$, $Z \rightarrow \tau\tau$) are important backgrounds to these searches and therefore a good understanding and description of these background processes is crucial. In particular, the $Z \rightarrow \tau\tau$ decay forms the dominant background to the Higgs boson search in the $H \rightarrow \tau\tau$ channel in the Standard Model as well as in the MSSM model. Since the $Z \rightarrow \tau\tau$ process has the same signature as the signal, a number of analysis techniques used in the $H \rightarrow \tau\tau$ analysis can be tested with the $Z \rightarrow \tau\tau$ process, e.g. mass reconstruction methods.

The measurement of the $Z \rightarrow \tau\tau$ cross section builds a complementary measurement to the precision measurements in the $Z \rightarrow ee$ and $Z \rightarrow \mu\mu$ channels at the LHC centre-of-mass energy. The measurements of the $Z \rightarrow \tau\tau$ cross section were performed by both ATLAS [2] and CMS [3] using data collected during 2010 corresponding to an integrated luminosity of 36 pb^{-1}. The new measurement performed with the ATLAS detector uses larger statistics of a data sample collected during 2011 [1]. The data sets correspond to an integrated luminosity of 1.55 fb^{-1} in the muon-hadron ($Z \rightarrow \tau\tau \rightarrow \mu + \text{hadrons} + 3\nu$, denoted as $\tau_\mu\tau_\text{h}$) and electron-muon ($Z \rightarrow \tau\tau \rightarrow e\mu + 4\nu$, denoted as $\tau_e\tau_\mu$) state and 1.34 fb^{-1} in the electron-hadron ($Z \rightarrow \tau\tau \rightarrow e + \text{hadrons} + 3\nu$, denoted as $\tau_e\tau_\text{h}$) final state.

The $Z \rightarrow \tau\tau$ process with one τ lepton decaying leptonically and the other one hadronically (denoted as $\tau_\ell\tau_\text{h}$) with a branching ratio of 45.6 % plays an important role in the hadronic τ reconstruction and identification studies. The advantage of the $\tau_\ell\tau_\text{h}$ channel is that a single lepton trigger can be used and therefore an unbiased sample of hadronic τ leptons can be selected. The variables used for the τ identification can be studied with this sample. Moreover, the measurement of the τ trigger and τ identification efficiency can be performed.

5.2 Signal and Background Processes

Details about the τ lepton decay modes and the typical $Z \rightarrow \tau\tau$ signature are presented in this section. The most relevant background processes are also discussed.

---

[1] The τ lepton reconstruction in ATLAS means the reconstruction of the visible part of the hadronically decaying τ lepton ($\tau \rightarrow \text{hadrons} + \nu_\tau$) where the visible part is built by the hadrons from the τ decay. More details about the τ lepton decays are given in Sect. 5.2.1 and the hadronic τ reconstruction is described in Sect. 5.3.2. The reconstructed hadronic τ lepton is called τ candidate in the following and denoted as τ_h.

5.2 Signal and Background Processes

5.2.1 $Z \rightarrow \tau\tau$ Signal Signature

The τ leptons are very unstable and have a short lifetime (mean lifetime $c\tau = 87\,\mu\text{m}$) [4]. That is the reason why the τ leptons decay before entering the detector and only their decay products can be detected. The τ leptons' decays can be divided into two categories:

- **leptonic decays** Decay modes with an electron $\tau^- \rightarrow e^- \bar{\nu}_e \nu_\tau$ (17.9 %) or a muon $\tau^- \rightarrow \mu^- \bar{\nu}_\mu \nu_\tau$ (17.4 %) in the final state belong to the leptonic τ decays. The reconstructed electron or muon in the detector plus missing transverse energy is the typical signature of the leptonically decaying τ in the detector.
- **hadronic decays** Most of the hadronic decays can be characterised by one or three charged pions and a tau neutrino possibly accompanied with few neutral pions. More rare are decays involving other mesons, e.g. kaons. The hadronic decays represent 64.7 % of τ leptons' decays. The hadronic decays are commonly categorised in groups characterised by the number of charged particles in the final state, i.e. by the number of tracks in the inner detector: Most frequent are decays with one charged particle (76.5 % of hadronic τ decays), so-called 1-prong decays. The decays with three tracks in the inner detector (23.4 % of hadronic τ decays), so-called 3-prong decays, are more rare.

The typical signature of the hadronic τ decays in the detector is a collimated jet with low track multiplicity and a relatively narrow energy deposition in the calorimeter compared to a jet produced by the hadronization of a quark or a gluon. Whenever a τ reconstruction in the ATLAS experiment is mentioned, the reconstruction of the hadronic part (the collimated jet) of the hadronically decaying τ lepton is meant.

The electron-hadron final state, $Z \rightarrow \tau\tau \rightarrow e + \text{hadrons} + 3\nu$, discussed in this thesis has a branching fraction of 23.1 %. The typical signature is an isolated lepton, a hadronic τ and a transverse missing energy coming from the neutrinos.

5.2.2 Background Processes

The hadronic τ reconstruction and identification is more difficult than in the case of leptons.[2] Moreover, a quark or gluon jet might be easily misidentified as a hadronic τ. Most of the background processes can be characterised by a true lepton and a jet faking the τ candidate.

The dominant background processes in the $\tau_\ell \tau_h$ channel are:

- **Multijets** The multijet background with its large production cross section has to be carefully taken under control. The τ candidate is typically a misidentified jet

---

[2] Lepton, denoted as ℓ, stands for an electron or a muon only in the following.

54   5 $Z \to \tau\tau$ Cross Section Measurement

in these processes. The lepton candidate can be either genuine (e.g. leptons from heavy-flavour decays) or a jet misidentified as a lepton (so-called fake lepton). The lepton candidate is situated inside a jet in most cases and therefore isolation criteria can be used to reduce the multijet background significantly.

- **W+jets** The W boson production is often accompanied by a jet which might be misidentified as a hadronic τ. Two different W decays can contribute to the background events: $W \to \ell\nu_\ell$ and $W \to \tau\nu_\tau \to \ell + 3\nu$. In most cases the lepton is a real lepton from the W boson decay, while the τ candidate is a jet misidentified as a hadronic τ. The W+jets event topology is different from the signal and the angular correlations between the missing transverse energy, the lepton and the τ candidate can be used to distinguish the signal from the W+jets background. The W+jets background is divided into $W \to e\nu$ and $W \to \tau\nu$ events in tables and figures for the $\tau_e\tau_h$ channel in this chapter.

- **Z+jets** Another important electroweak background comes from the decays of the Z boson to leptons ($Z \to \ell\ell$), possibly accompanied with one or more jets. Typically, one lepton from the Z boson decay is correctly measured and identified. The fake hadronic τ might come either from the second lepton or from the jet in the event. A veto against events with two or more leptons are applied to suppress the $\gamma^*/Z \to \ell\ell$ background.
  The so-called Z+jets background is denoted as $\gamma^*/Z \to ee$ in tables and figures for the $\tau_e\tau_h$ channel in this chapter.

- **$t\bar{t}$** The contribution of the top background ($t\bar{t} \to WWbb$) is rather small compared to the other background sources and no special cut against this background is applied.

- **Dibosons** The cross section of the dibosons production (WW, ZZ, WZ) is very small and the contribution from this background is not very significant in the $Z \to \tau\tau$ analysis. Nevertheless, it is taken into account.

5.2.3 Data and Monte Carlo Samples

Data Collection

The data samples used in the analysis were collected during 2011 data taking period. Only a good collision data at $\sqrt{s} = 7$ TeV with stable beam conditions are used. The good quality requirement selects only luminosity blocks from all data for which no serious defects in the various subsystems are reported. These quality criteria are analysis dependent, for various analysis different subsystems and physical objects are of the main interest and therefore different data quality requirements (so-called Good Run Lists) are relevant.

Only data collected during a first part of 2011 data taking period are analysed. The main reason why only a part of the data collected during 2011 is used is the trigger stability. The combined electron and tau trigger is utilised in the $\tau_e\tau_h$ analysis and

5.2 Signal and Background Processes 55

the isolated muon trigger in the other two channels (more details about the trigger selection are given in Sect. 5.3.1). The requirement of the stable setup of these triggers restricts the selection of data samples to the integrated luminosity of 1.34 fb^{-1} in the $\tau_e\tau_h$ and 1.55 fb^{-1} in the $\tau_\mu\tau_h$ and $\tau_e\tau_\mu$ after the data quality checks.

Monte Carlo Samples

The Monte Carlo samples for the signal and background used in the analysis are generated at $\sqrt{s} = 7$ TeV and are passed through the full detector simulation based on the Geant4 simulation programme [5]. The electroweak decays of W and γ^*/Z, for both signal and background, are generated using the Alpgen [6] generator, interfaced to HERWIG [7] and JIMMY [8], with CTEQ6L1 [9] parton distribution function (PDF) and are normalised to NNLO cross sections [10–12]. MC@NLO generator [13] is used for the $t\bar{t}$ background and HERWIG generator for diboson samples. The τ leptons' decays are performed with TAUOLA [14] where the spin correlations are correctly modelled. All generators are interfaced to PHOTOS [15] where the effect of the final state QED radiation is simulated.

Pile-up in Data and in Monte Carlo

Data samples collected during 2011 suffer from relatively high pile-up contribution. The average number of minimum bias collisions per event is approximately 6 in the analysed period of 2011 [16]. Due to the small separation between individual proton bunches at the LHC (50 ns in the analysed data), two different pile-up features are observed: Out-of-time pile-up (influence from interactions from previous bunch crossings) and in-time pile-up (interactions from the same bunch crossing). More details about the pile-up, with the emphasis on the pile-up contribution in the hadronic calorimeter TileCal, can be found in Sect. 3.3.

The pile-up contribution is also simulated in the Monte Carlo samples where the appropriate number of minimum bias collisions is added on top of the simulated hard-scattering process during the digitisation procedure. However, the Monte Carlo samples used have been generated before the conditions of 2011 data taking were known. That is the reason why the simulated pile-up conditions are different from the real data conditions as shown in Fig. 5.1. Therefore the simulated events are re-weighted in a way that the average number of interactions per bunch crossing agrees with data after the re-weighting procedure.

5.3 Object and Event Selection

The object and event selection with the emphasis on the $\tau_e\tau_h$ channel is summarised below. The $\tau_\mu\tau_h$ and $\tau_e\tau_\mu$ selection is mentioned only briefly and more details can be found in Ref. [1].

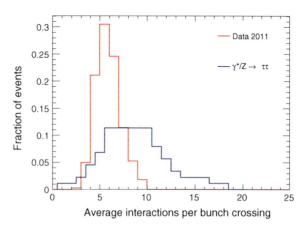

Fig. 5.1 Comparison between average number of pile-up collisions in data and signal Monte Carlo before the pile-up re-weighting is applied on the simulated sample in the $\tau_e\tau_h$ channel

5.3.1 Event Preselection

Event Cleaning

Good collision events are selected by requiring at least one primary vertex with four or more associated tracks. Moreover, the event is rejected if there might be a jet or a τ candidate caused by cosmic-ray events or by known noise effects in the calorimeter. Furthermore, the LAr calorimeter experienced hardware problems during the 2011 data taking period which has resulted in an acceptance hole in the calorimeter. The reconstructed objects (jets, hadronic τ or electrons) are rejected if they are localised in the region of the readout problems in the electromagnetic calorimeter. The acceptance hole is not simulated in the Monte Carlo samples, but a correction accounting for this acceptance loss is applied to the simulated samples.

Trigger Requirement

Triggers with lowest possible threshold which remains unprescaled during the considered data taking period in 2011 are used in the $Z \rightarrow \tau\tau$ analysis.

Single muon triggers are used in the $\tau_\mu\tau_h$ and $\tau_e\tau_\mu$ channels. The triggers EF_mu15i and EF_mu15i_medium search for a muon candidate with transverse momentum higher than 15 GeV and a requirement of a loose muon isolation is applied at the event filter (EF) level.

The combined electron and hadronic tau trigger, EF_tau16_loose_e15_medium, is used in the $\tau_e\tau_h$ analysis. The combined trigger requires an electron candidate with $E_T > 15$ GeV together with a τ candidate with $E_T > 16$ GeV, both passing specific identification criteria.

5.3 Object and Event Selection

Due to the changes in the trigger setup during the 2011 data taking period, different trigger hypothesis for the hadronic τ part of the trigger is available in the Monte Carlo and in the data samples. On top of that, it was not possible to emulate the τ trigger decision with the Monte Carlo samples being used. Therefore the trigger decision of the combined trigger is considered only in the data samples and a special treatment of the trigger is used in the Monte Carlo samples. Only the electron trigger (EF_e15_medium) decision is taken into account in the simulated samples and the efficiency of the τ trigger part (EF_tau16_loose) is used as an event weight instead of the trigger decision. The τ trigger efficiencies have been measured in bins of the τ candidate E_T by means of the $Z \to \tau\tau$ tag and probe method. This treatment does not described all features of the trigger selection, but it is applicable for the cross section measurement.

5.3.2 Object Preselection and Selection

Several offline reconstructed objects (electrons, muons, τ candidates, jets, missing transverse energy) enter the analysis. The object selection is done in several steps: First, looser selection criteria are required for the event preselection. The preselected objects are used for the removal of the overlapping objects (so-called overlap removal) and also in the dilepton veto against the $Z \to \ell\ell$ background. Tighter selection cuts are applied on the leptons and the hadronic τ in the next step. Finally, isolation of the lepton candidates is required. The leading isolated lepton passing the tight selection cuts and the selected τ candidate are used further in the analysis for calculation of derived quantities that are used for reduction of the background contamination (Sect. 5.3.3).

Electrons

The electrons play an important role in the $\tau_e\tau_h$ and $\tau_e\tau_\mu$ channels as a real electron is present in both final states. Strict identification criteria are applied on the reconstructed electron to select a clean sample of electrons.

Details about the electron reconstruction and identification can be found in Ref. [17]. The electron reconstruction algorithm (described in Sect. 4.1.1) looks for a cluster in the electromagnetic calorimeter with an associated track in the inner detector. Afterwards, electron identification criteria (discussed in Sect. 4.1.2) are applied on the reconstructed electrons in order to enhance the purity of the selected electron sample.

Preselection Electrons passing so-called medium identification criteria with transverse energy $E_T > 15$ GeV within the pseudorapidity range $|\eta| < 2.47$ excluding the transition region between the barrel and end-cap calorimeters ($1.37 < |\eta| < 1.52$) are preselected. The quality of the electron candidates is checked and only "good" electrons are accepted.

58   5 $Z \to \tau\tau$ Cross Section Measurement

Selection A higher transverse energy cut and stricter electron identification are required for the selected electrons. The transverse energy threshold is raised to 17 GeV to avoid the region close to the electron trigger threshold (15 GeV at EF) which is difficult to model well in the Monte Carlo simulations. Only electrons passing the highest level of the electron identification criteria, so-called tight identification cuts, are accepted.

Muons

The preselected muons are important for removing τ candidates that might be reconstructed from real muons. The selected muons are not used in the $\tau_e\tau_h$ channel, but they are used in the $\tau_\mu\tau_h$ and $\tau_e\tau_\mu$ channels and mentioned here for completeness. More details about the muons reconstruction in ATLAS can be found in Ref. [18].

Preselection Muons with transverse momentum $p_T > 6$ GeV in the pseudorapidity region $|\eta| < 2.7$ are preselected if they pass "loose" identification. These muons include muons reconstructed as a combination of the measurements in the inner detector and the muon spectrometer as well as stand alone muons reconstructed in the muon chambers only. The loose selection is important for the removal of hadronic τ candidates overlapping with muons.

Selection Only muons that are built as a combination of the inner detector and the muon spectrometer measurements are accepted in the selection. The transverse momentum cut is raised to 17 GeV and the pseudorapidity range is reduced to $\eta < 2.4$ corresponding to the trigger acceptance. In order to remove muons not originating in the collision, the longitudinal impact parameter is required to be less than 10 mm with respect to the primary vertex. A number of track quality cuts (based on the number of hits in different sub-detectors of the inner detector) are applied on the muon candidate track to reduce the contribution of fake muons.

Jets

The reconstructed jets are used as the seeds for the hadronic τ reconstruction. The jets are reconstructed with an anti-k_t algorithm [19, 20] with a distance parameter $R = 0.4$ with three-dimensional topological clusters built from the calorimeter cells.

Jets also enter the event cleaning procedure in the $\tau_\ell\tau_h$ channel where jets with transverse energy larger than 20 GeV within the pseudorapidity range $|\eta| < 4.5$ are used.

Hadronic τ Candidates

Selecting a clean sample of hadronic τ is crucial for the $\tau_\ell\tau_h$ channels. Strict identification criteria are applied to reduce the multijet background where a jet might be misidentified as a τ candidate. In addition, a veto against electrons that might also

5.3 Object and Event Selection

be faking the hadronic τ is considered. Details about the hadronic τ reconstruction and identification in ATLAS can be found in Ref. [21].

Calorimeter jets, reconstructed as described above, with transverse momentum $p_T > 10$ GeV form seeds for the hadronic τ reconstruction. Inner detector tracks with $p_T > 1$ GeV passing dedicated track quality cuts are associated to the τ candidates. The identification variables based on the tracking and calorimeter information are derived (e.g. invariant mass of the associated tracks, E_T over p_T of the leading track, fraction of E_T within $\Delta R < 0.1$ of the τ candidate, fraction of E_T of the τ candidate in the EM calorimeter) and used to distinguish between a jet, resp. an electron and a τ candidate.

Preselection The τ candidate is required to be in the pseudorapidity range $|\eta| < 2.47$ excluding the crack region ($1.37 < |\eta| < 1.52$). The minimal transverse momentum cut is set to 20 GeV, resp. 25 GeV in the $\tau_\mu \tau_h$, resp. $\tau_e \tau_h$ channel. The higher threshold in the $\tau_e \tau_h$ channel is necessary to avoid the region on the turn-on curve of the tau trigger efficiency. In addition, the τ candidates with a leading track within $|\eta| < 0.03$ are excluded due to a high fake rate from electrons in this region caused by the gap in the calorimeter acceptance and the reduced TRT coverage around $|\eta| = 0$. No τ identification is performed at the preselection step.

Selection The identification requirements on the τ candidate are applied during the object selection. The separation between the jet and the τ candidate is done with a Boosted Decision Tree (BDT) method described in Ref. [21]. The BDT is trained to define three different working points (*loose, medium* and *tight*) with increasing background rejection and decreasing signal efficiency. In order to have a flat signal efficiency, the cut on the resulting BDT score depends on the τ candidate's transverse energy. The signal efficiency for the BDT medium selection, which is found to be optimal for the $Z \rightarrow \tau\tau$ analysis, is approximately 45 %. Not only jets can fake hadronic τ leptons, but also electrons can form a fake τ candidate. A cut-based electron veto [21] with the strongest rejection, so-called tight veto, is applied to select a sample of good τ candidates.

Missing Transverse Energy

A non-zero missing transverse energy (E_T^{miss}) coming from the neutrinos in the τ lepton decays is characteristic for the signal events. However, no direct cut on the missing transverse energy is applied, but the missing energy is used in the derived variables to suppress the W+jets background as described in Sect. 5.3.3.

The reconstruction of the missing transverse energy starts from the energy deposits in the calorimeter clusters and the reconstructed muon tracks and the correction accounting for the muons' energy lost in the calorimeters is considered as described in Sect. 4.3.2.

Overlap Removal

Multiple candidates (electrons, muons, τ candidates or jets) might be reconstructed from the same localised response in the ATLAS detector. An unique hypothesis for each object is required in the $Z \to \tau\tau$ analysis. Therefore an overlap removal procedure is performed.

Since muons and electrons are selected with a higher purity than hadronic τ leptons, any preselected τ candidate is removed from the consideration if it lies within $\Delta R < 0.4$ from any preselected lepton. The $Z \to \ell\ell$ background with one lepton being mismeasured as a hadronic τ candidate is suppressed by this requirement.

In the next step, the muon objects are considered being more pure than electrons and they are taken with higher priority. Accordingly, an electron candidate is removed if it overlaps with a muon within $\Delta R < 0.2$.

Lepton Isolation

The leptons from the $Z \to \tau\tau$ decay are preferably isolated. On the contrary, the electrons and muons observed in the multijet events (e.g. muons coming from B-hadron decays) do not tend to be isolated. Consequently, requiring an isolated lepton is an efficient way to reduce the huge multijet background.

Two kinds of isolation criteria are used in the analysis, the track and the calorimeter isolation. Although they are both defined in Sect. 4.5, a short reminder is added also here: The track isolation variable is calculated as a scalar sum of the transverse momentum of charged particles' tracks in a cone of radius $\Delta R = 0.4$ in the $\eta - \phi$ space excluding the track belonging to the lepton candidate itself (denoted as $p_{\mathrm{T}}\mathrm{Cone}40$). The calorimeter isolation is defined in a similar way, the transverse energy in the electromagnetic calorimeter in a given cone is summed up excluding the calorimeter deposits associated to the lepton itself (denoted as $E_{\mathrm{T}}\mathrm{Cone}40$ for radius $\Delta R = 0.4$). In order to minimise the dependence of the isolation variables on the lepton's momentum, the relative isolation (isolation variable divided by the transverse momentum, resp. energy of the muon, resp. electron candidate) is used. Moreover, the calorimeter isolation for the electrons was found to be dependent on pile-up. Therefore a special correction is applied to make the isolation variable more robust against the pile-up.

The isolation cuts are optimised to reduce the large multijet background while not rejecting too large fraction of signal events. The cut values are determined by studying the signal and background efficiencies. The signal efficiency is estimated from the Monte Carlo simulations whereas the multijet background contribution is studied in data. A multijet-rich sample is constructed by requiring the selected electron and the selected tau candidate to have charges of the same sign to enhance the multijet fraction. Contributions of electroweak backgrounds (W, Z) and $t\bar{t}$ are subtracted from the measurements in data to extract the multijet efficiency. The distribution of the electron calorimetric and track isolation for signal and background events are shown in Fig. 5.2.

5.3 Object and Event Selection

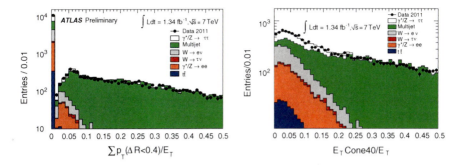

Fig. 5.2 Comparison between electron isolation variables in data and in Monte Carlo simulations after selection of an electron and a τ candidate with opposite charges in the $\tau_e\tau_h$ channel. Most signal events are found in the first bin of the track isolation (*left hand side*) plot) corresponding to events with no additional track in the cone with $\Delta R = 0.4$. The rest of the signal events falls in the few nearby bins

The following isolation cuts are applied on the selected electrons and muons:

- Electrons: $p_T\text{Cone}40/E_T < 0.06$ and $E_T\text{Cone}40/E_T < 0.1$
- Muons: $p_T\text{Cone}40/p_T < 0.03$ and $E_T\text{Cone}30/p_T < 0.04$

The stricter isolation criteria for muons are motivated by the usage of the isolation already at the trigger level. The offline selection has to be tighter to enable the multijet background estimation as described in Sect. 5.4.3.

Lepton's Efficiency Correction Factors

The lepton's reconstruction, identification and trigger efficiencies, as well as efficiencies of the isolation criteria applied on the selected lepton have been measured by means of the tag and probe method in both data and in Monte Carlo simulations as described in Sect. 4.2. The details about the measurement of the electron reconstruction efficiency can be found in Ref. [17], whereas the other measurements of the electron efficiency are discussed in Chap. 4 (identification efficiency in Sect. 4.3, single electron trigger efficiency in Sect. 4.4 and isolation cuts efficiency in Sect. 4.5). The individual correction factors (so-called scale factors), ratios between efficiency measured in data and in Monte Carlo, have been derived and they are applied as an event weights in the simulations to obtain a good agreement between simulated and real data samples.

5.3.3 Event Selection in the $\tau_\ell\tau_h$ Channel

The selection of the τ candidate and the isolated lepton is the first step in the event selection. The kinematic distributions (transverse energy and pseudorapidity) of the selected τ candidates and the isolated lepton are shown in Fig. 5.3. The lepton and the hadronic τ are required to have opposite sign charges in order to be able to estimate the multijet background contribution. The estimation of different backgrounds is described in Sect. 5.4. The largest background contribution comes from W+jets events, followed by Z+jets and multijet background at this step of the cutflow selection. Further event selection criteria are motivated by the suppression of the electroweak and multijet backgrounds as described in this section.

Opposite Sign Requirement

Signal events can be characterised by opposite charges of the lepton and the τ candidate. The τ candidate charge (Q_τ), which is reconstructed as the sum of charges of the associated tracks, is required to be of the opposite sign with respect to the charge of the lepton (Q_ℓ): $Q_\ell \cdot Q_\tau < 0$.

Multijet background events where a jet fakes the hadronic τ do not prefer the opposite sign of the lepton and τ candidate charges. The opposite sign cut is used to reduce further this background contamination.

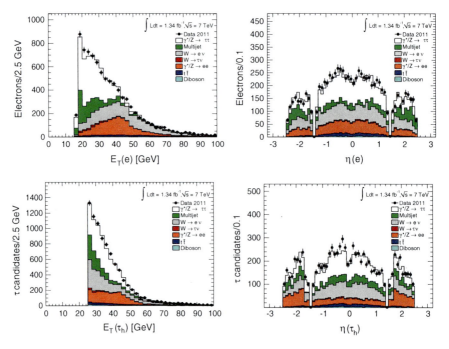

Fig. 5.3 The distribution of E_T and η of the selected and isolated electron (*upper row*) and the selected τ candidate (*lower row*) after the object selection step with an additional requirement of the opposite charges of the lepton and the τ candidate in the $\tau_e\tau_h$ channel

5.3 Object and Event Selection

Dilepton Veto

The $Z \to \ell\ell$+jets background with a jet misidentified as a τ candidate is suppressed by removing events where a second lepton is found. The preselected leptons are used to enhance the power of the dilepton veto.

Not only events with two or more same flavor leptons are vetoed, but the event is also not considered if an electron and a muon occur in one event. This requirement reduces the contribution of $Z \to \tau\tau \to e\mu + 4\nu$ decay which is analysed as a separate channel.

The number of preselected leptons is shown in Fig. 5.4. The event is vetoed if more than one preselected lepton (electron or muon) is reconstructed.

Cuts Against W+jets Background

The W+jets background can contribute in two W decay modes: $W \to \ell\nu_\ell$ and $W \to \tau\nu_\tau \to \ell + 3\nu$. In most cases the jet is badly identified as the τ candidate and the lepton is a real lepton from the W decay. Due to the fact that the W+jets background has a different topology than $Z \to \tau\tau$ signal, the W+jets can be suppressed to a large extent.

The Z boson mass is much larger than the τ lepton's mass and thus the τ leptons are boosted with their decay products being collimated along the τ lepton direction. The missing transverse energy in the $Z \to \tau\tau$ decays is formed by the neutrinos' energy. In most cases the Z boson is born with low transverse momentum and therefore the τ leptons tend to be produced back-to-back in the transverse plane. If the Z boson has a larger transverse momentum, then the E_T^{miss} vector is located within the angle formed by the visible Z boson decay products. On the contrary, the decay products

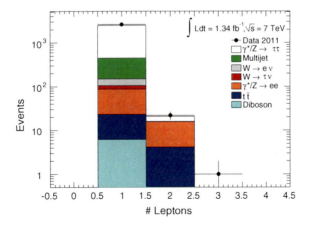

Fig. 5.4 Number of preselected leptons for events passing all event selection cuts except the dilepton veto in the $\tau_e \tau_h$ channel

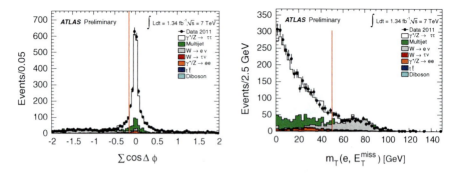

Fig. 5.5 Variables used for W+jets background suppression in the $\tau_e\tau_h$ channel. The vertical *red line* indicates the cut value used in the event selection. The plots are shown for events passing all event selection cuts except the two W+jets cuts

from $W \to \ell\nu_\ell$+jets event (the electron, the jet misidentified as the hadronic τ and the neutrino) are distributed in a way to build a p_T balance in the transverse plane. That is the reason why the missing transverse energy points outside the angle between the fake τ candidate and the lepton in most cases. Similar angular correlations occur in the $W \to \tau\nu_\tau \to \ell + 3\nu$+jets events.

Two variables are built in order to reduce the W+jets background. The first one is defined as follows

$$\sum \cos\Delta\phi = \cos(\phi(\ell) - \phi(E_T^{\text{miss}})) + \cos(\phi(\tau_h) - \phi(E_T^{\text{miss}})). \quad (5.1)$$

The distribution of this variable is shown on the *left hand side* in Fig. 5.5. Most of the signal events are localised in the peak around zero which corresponds to the case where the decay products are produced back-to-back in the transverse plane. In addition, the $Z \to \tau\tau$ events have a tail into positive $\sum \cos\Delta\phi$ values which are characterised by the E_T^{miss} vector pointing inside the angle between the τ candidate and the lepton. On the other hand, the W+jets events tend to the negative $\sum \cos\Delta\phi$ values corresponding to the E_T^{miss} vector pointing outside this angle. The events with $\sum \cos\Delta\phi > -0.15$ are considered for further analysis.

The second variable used against the W+jets background is the transverse mass of the lepton and missing transverse energy as defined in Eq. (4.2). The transverse mass distribution for the signal and background events is shown in Fig. 5.5, *Right*. The $Z \to \tau\tau$ tends to the low values of the transverse mass. On the contrary, the transverse mass distribution prefers larger values in the W+jets events. The events are accepted only if $m_T < 50$ GeV is fulfilled.

5.3 Object and Event Selection

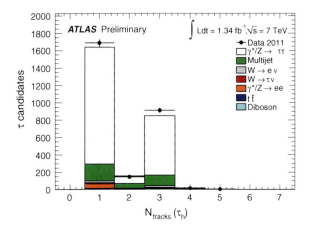

Fig. 5.6 Number of tracks associated to the τ candidate for events passing all event selection cuts except hadronic τ candidate's cleaning cuts in the $\tau_e\tau_h$ channel

Further Requirements on the Hadronic τ Candidate

More requirements on the τ candidates are applied during the event selection to further reduce fake τ candidates coming from badly identified QCD jets.

The τ candidates are required to have exactly one or three associated tracks measured in the inner detector. The distribution of the number of τ candidates' tracks is shown in Fig. 5.6. A small amount of signal events falls in the 2-track bin. Approximately half of these are 3-prong τ leptons with one of the tracks not reconstructed, while the rest are 1-prong τ leptons with an additional close-by track.

Moreover, the τ lepton charge is required to be ± 1 where the charge is calculated as the sum of charges of the associated tracks.

Visible Mass Window

In order to increase the purity of the $Z \to \tau\tau$ signal events and minimise the contamination from the $Z \to \ell\ell$ events, a cut on the so-called visible mass of the τ candidate and the lepton is applied. The visible mass is defined as the invariant mass of the lepton and the hadronic τ candidate (the neutrinos are not considered in the calculation)

$$m_{\text{vis}} = \sqrt{2 p_T(\ell) \cdot p_T(\tau_h) \cdot [\cosh(\eta(\ell) - \eta(\tau_h)) - \cos(\phi(\ell) - \phi(\tau_h))]}. \quad (5.2)$$

The distribution of the visible mass is shown in Fig. 5.7. While the $Z \to \ell\ell$ events are expected to have a maximum in the region of $m_{\text{inv}} \sim 90$ GeV, the $Z \to \tau\tau$ signal tends to lower m_{vis} values with a peak around 60 GeV due to the missing energy of the neutrinos in the decay.

The selected events are required to be within the visible mass window 35–75 GeV.

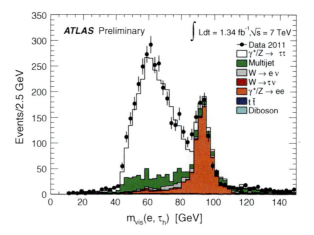

Fig. 5.7 Visible mass distribution for events passing all event selection cuts except the cut on the visible mass itself in the $\tau_e\tau_h$ channel

Table 5.1 Cutflow table for data, signal and background events in the $\tau_e\tau_h$ channel

	Data 2011	$Z/\gamma^* \to \tau\tau$	$W \to e\nu$	$W \to \tau\nu$
Object selection	15200 ± 123	3393 ± 33	4660 ± 57	291 ± 12
Opposite sign	8675 ± 93	3087 ± 32	2158 ± 39	127 ± 7
Dilepton veto	8441 ± 92	3067 ± 31	2149 ± 39	127 ± 7
W cuts	4649 ± 68	2570 ± 28	210 ± 12	50 ± 4
$N_{tracks}(\tau_h) = 1$ or 3	4358 ± 66	2456 ± 28	180 ± 11	41 ± 4
$|charge(\tau_h)| = 1$	4351 ± 66	2453 ± 28	179 ± 11	41 ± 4
$m_{vis} = 35-75$ GeV	2600 ± 51	2029 ± 25	45 ± 5	18 ± 2
	$Z/\gamma^* \to ee$	$t\bar{t}$	Dibosons	Multijets
Object selection	2362 ± 28	534 ± 4	174 ± 5	–
Opposite sign	1575 ± 24	340 ± 3	103 ± 4	1156 ± 60
Dilepton veto	1450 ± 24	271 ± 3	97 ± 4	1154 ± 58
W cuts	900 ± 19	59 ± 1	18 ± 2	726 ± 36
$N_{tracks}(\tau_h) = 1$ or 3	879 ± 19	54 ± 1	16 ± 1	593 ± 33
$|charge(\tau_h)| = 1$	878 ± 19	53 ± 1	16 ± 1	584 ± 32
$m_{vis} = 35-75$ GeV	64 ± 4	17 ± 1	6 ± 1	300 ± 21

The statistical uncertainties are given in the table. The way how the background contribution is estimated is described in Sect. 5.4

Summary of the Selection Cuts

The basic event selection cuts together with the event yields for data, signal Monte Carlo and the main background processes can be found Table 5.1 for the $\tau_e\tau_h$ channel. The largest background after the full selection comes from multijets. The way how different backgrounds are estimated is described in Sect. 5.4.

A number kinematic variables are shown for events passing all event selection cuts in Figs. 5.8 and 5.9. Namely, the selected τ candidate's and lepton's transverse

5.3 Object and Event Selection

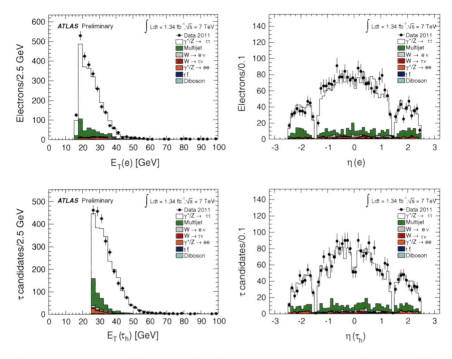

Fig. 5.8 The distribution of E_T and η of the isolated electron (*upper row*) and selected τ candidate (*lower row*) after all selection cuts are applied in the $\tau_e\tau_h$ channel

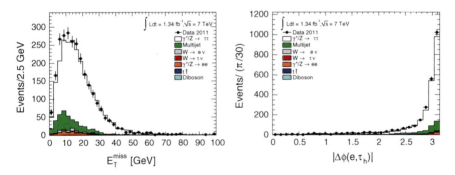

Fig. 5.9 The distribution of E_T^{miss} and the absolute value of $\Delta\phi$ between the τ candidate and the lepton after all selection cuts are applied in the $\tau_e\tau_h$ channel

energy and pseudorapidity, the missing transverse energy and the absolute value of $\Delta\phi$ between the τ candidate and the lepton. An overall acceptable agreement between data and Monte Carlo prediction can be observed.

5.4 Background Estimation

The diboson and $t\bar{t}$ backgrounds contribution is very small after all selection cuts are applied. The event yields from these backgrounds are entirely estimated from the Monte Carlo simulations. The individual Monte Carlo samples are normalised to the required integrated luminosity $(\mathcal{L})$ using the theoretical cross sections (σ) where the number of events (N) is given by $N = \mathcal{L} \cdot \sigma$.

Differences between the Monte Carlo predictions and collision data were observed in processes with a jet being misidentified as a τ candidate. Consequently W+jets and Z+jets contributions cannot be taken directly from the Monte Carlo simulations, but a normalisation factor is to be derived from data in W boson, resp. Z boson rich control region. The procedure is described in Sects. 5.4.1, 5.4.2.

Due to the fact that the multijet background cross section is several orders of magnitude higher than the electroweak processes, the statistics available in the Monte Carlo simulations is not sufficient to provide reliable predictions. Moreover, the jet-τ fake rate efficiency in Monte Carlo simulations differs from the measurements in data. The estimation of the multijet background is done completely by data-driven method as described in Sect. 5.4.3.

5.4.1 W+jets Background Normalisation

It has been observed that the number of W+jets background events agree reasonably well in data and in Monte Carlo simulations before the hadronic τ identification is applied. However, after requiring the τ candidate to pass the identification criteria, the event yield predicted by Monte Carlo is higher than the actual number of W+jets events measured in data. In other words, the Monte Carlo overestimates the number of QCD jets being misidentified as a τ candidate after the τ identification criteria are applied.

The comparison between data and Monte Carlo simulations is performed in a W-enriched control region. This region is defined as follows: The selected τ candidate and the isolated lepton are required, the dilepton veto is applied and further requirements on the τ candidate (number of associated tracks, unit charge) are considered, but both W boson suppression cuts are inverted ($\sum \cos \Delta\phi < -0.15$ and $m_T > 50\,\text{GeV}$). The difference between the number of W events in data and predicted in Monte Carlo simulations in the W control region is shown on the *left hand side* in Fig. 5.10.

The Monte Carlo predictions for both $W \rightarrow \ell\nu$ and $W \rightarrow \tau\nu$ event yields are scaled by a normalisation factor k_W. The factor k_W is defined as a ratio of events measured in data in the W control region subtracting the small contamination from other backgrounds ($Z \rightarrow \ell\ell$, $t\bar{t}$ and dibosons) and the number of W events predicted in the Monte Carlo simulations. The contribution of the $Z \rightarrow \ell\ell$, $t\bar{t}$ and diboson backgrounds in the W control region is taken from the Monte Carlo simulations.

5.4 Background Estimation

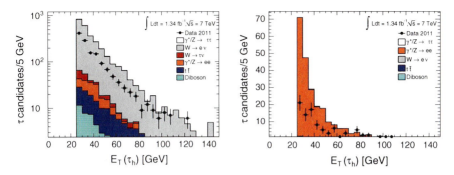

Fig. 5.10 Transverse momentum distribution of the τ candidate in the W control region on the *left* and Z control region on the *right* in the $\tau_e\tau_h$ channel. The Monte Carlo simulations overestimate the event yield compared to the measurements in data and a normalisation factor is to be applied on the Monte Carlo predictions. The signal ($Z \to \tau\tau$) contamination in the W and Z control regions is found to be negligible

$$N_W^{\text{WCR}} \to k_W N_W^{\text{WCR}} = N_{\text{data}}^{\text{WCR}} - N_{Z\to\ell\ell}^{\text{WCR}} - N_{t\bar{t}}^{\text{WCR}} - N_{\text{diboson}}^{\text{WCR}}. \quad (5.3)$$

The measurement of the k_W factor is provided for two different cases determined by the charge product of the τ candidate and the lepton. The motivation for that is that the τ misidentification rate is different for jets coming from a quark or a gluon hadronization and thus a different value of the normalisation factor is expected. The W+quark process prefers opposite sign charges, while there is no such expectation in the W+gluon process. The measurement for the same sign case is necessary for the multijet background estimation as discussed in Sect. 5.4.3.

The measured k_W factor with its statistical uncertainty in the $\tau_e\tau_h$ channel is

- Opposite sign case: $k_W = 0.44 \pm 0.02$ (stat.)
- Same sign case: $ k_W = 0.56 \pm 0.04$ (stat.)

5.4.2 Z+jets Background Normalisation

The Z background contributes in two different ways depending whether a lepton or a jet is misidentified as a τ candidate. Since the probability for an electron to be misidentified as a hadronic τ is higher than for a muon, events with one of the leptons misidentified as a τ candidate are more frequent in the $\tau_e\tau_h$ channel than in the $\tau_\mu\tau_h$ channel. Scale factors derived from a $Z \to ee$ tag and probe study [21] are used to correct the electron misidentification probability in the Monte Carlo simulations in the $\tau_e\tau_h$ channel.

The second case where the jet is misidentified as the hadronic τ suffers from a similar normalisation problem as described for the W+jets background. A normalisation factor k_Z is found in the Z-enriched control region. The events in the Z control region are defined by a requirement of two reconstructed same-flavor leptons with the invariant mass in the region close to the nominal Z boson mass ($66 < m_{\ell\ell} < 116\,\mathrm{GeV}$) along with the selected τ candidate. The Z control region is very pure as can be seen on the *right hand side* in Fig. 5.10 and no background subtraction is necessary.

The k_Z factor is applied to $Z \rightarrow \ell\ell$ Monte Carlo simulations only in events where a jet is misidentified as a hadronic τ, i.e. the hadronic τ is not matched to a lepton at the truth level. The measured k_z factor with its statistical uncertainty in the $\tau_e\tau_h$ channel is

- $k_Z = 0.39 \pm 0.05$ (stat.)

5.4.3 Multijet Background

The multijet background is suppressed to a large extent during the event selection, but it remains the dominant background in the $\tau_\ell\tau_h$ channel. The so-called *ABCD* method is used to estimate the number of multijet background at different steps of the event selection. Four statistically independent regions with different requirements on the lepton isolation (passing or failing the isolation criteria) and on the charge product of the τ candidate and the lepton (opposite or same sign) are defined. The opposite sign to same sign ratio (R_{OSSS}), which is assumed to be independent on the lepton isolation, is evaluated in the QCD-rich regions with non-isolated leptons. Then the number of multijet events in the signal region is extrapolated from the region with isolated lepton and same sign requirement using the R_{OSSS} ratio.

In more details, the regions are defined as follow:

- Region A: the signal region with the isolated lepton and the opposite sign requirement.
- Region B: the control region with the isolated lepton and the same sign requirement.
- Region C: the control region with the non-isolated lepton and the opposite sign requirement.
- Region D: the control region with the non-isolated lepton and the same sign requirement.

The number of multijet events in the control regions B, C, D is evaluated after the subtraction of the electroweak and $t\bar{t}$ processes which is estimated from the Monte Carlo simulations (k_W, resp. k_Z factor is applied to W boson, resp. $Z \rightarrow \ell\ell$ background)

$$N^i_{\mathrm{Multijet}} = N^i_{\mathrm{data}} - N^i_{Z\rightarrow\tau\tau} - N^i_{Z\rightarrow\ell\ell} - N^i_{W\rightarrow\ell\nu} - N^i_{W\rightarrow\tau\nu} - N^i_{t\bar{t}} - N^i_{\mathrm{diboson}} \quad (5.4)$$

where $i = B, C, D$.

The opposite sign to same sign ratio (R_{OSSS}) is evaluated as a ratio of the number of multijet events in regions C and D ($R_{OSSS} = N^C_{\mathrm{Multijet}}/N^D_{\mathrm{Multijet}}$). The regions

5.4 Background Estimation

Table 5.2 Number of events in regions B, C, D used for the multijet background estimation in the $\tau_e\tau_h$ channel

Regions	B	C	D
Data	353	2626	2403
$\gamma^*/Z \to \tau\tau$	19	71	4
$\gamma^*/Z \to \ell\ell$	29	3	–
$W \to \ell\nu$	15	2	–
$W \to \tau\nu$	5	1	–
$t\bar{t}$	2	3	1
Diboson	1	–	–
Multijet	282	2546	2397

C and D are very pure in multijet events as can be seen in Table 5.2. The multijet background estimation in the signal region is calculated using this equation

$$N^A_{\text{Multijet}} = \frac{N^C_{\text{Multijet}}}{N^D_{\text{Multijet}}} N^B_{\text{Multijet}} = R_{osss} N^B_{\text{Multijet}} . \tag{5.5}$$

The measured R_{OSSS} ratio with its statistical uncertainty in the $\tau_e\tau_h$ channel is

- $R_{OSSS} = 1.06 \pm 0.03$ (stat.)

The ratio is close to unity as expected.

5.4.4 Expected Number of Signal and Background Events

The expected numbers of signal and background events in the $\tau_e\tau_h$ channel corresponding to the integrated luminosity of 1.34 fb^{-1} are summarised in Table 5.3. The estimated number of events for different background processes are derived as described in this section. Furthermore, the total number of data after the full selection procedure (N_{obs}) and the expected number of signal events based on the Monte Carlo simulations are given.

5.5 Methodology for Cross Section Calculation

The cross section $\sigma(Z \to \tau\tau)$ within the $\tau\tau$ invariant mass window from 66 to 116 GeV is measured in each final state ($\tau_e\tau_h$, $\tau_\mu\tau_h$ and $\tau_e\tau_\mu$) separately. The measurement is performed as described in Ref. [2], using the formula

$$\sigma(Z \to \tau\tau) \times BR = \frac{N_{obs} - N_{bkg}}{A_Z \cdot C_Z \cdot \mathcal{L}} \tag{5.6}$$

where

- BR is the branching ratio for the considered final state, e.g. $BR(\tau \to e\nu\nu, \tau \to \tau_h\nu)$ in the $\tau_e\tau_h$ channel.
- N_{obs} is the number of observed events in data.
- N_{bkg} is the number of estimated background events. The way how the number of the background events is extracted is described in Sect. 5.4 and the number of background events is summarised in Table 5.3 for the $\tau_e\tau_h$ channel.
- A_Z is the kinematic and geometric acceptance for the signal process, more details are given below.
- C_Z is the correction factor which accounts for the efficiency of triggering, reconstructing and identifying decays within the geometrical acceptance. More details can be found below.
- $\mathcal{L}$ denotes the integrated luminosity.

So-called fiducial regions are used in the definition of the acceptance factor A_Z and the correction factor C_Z. The fiducial region is defined in this way in the $\tau_e\tau_h$ channel

- Electron: $E_T > 17$ GeV, $|\eta| < 2.47$, excluding $1.37 < |\eta| < 1.52$
- Tau: $E_T > 25$ GeV, $|\eta| < 2.47$, excluding $1.37 < |\eta| < 1.52$
- Event: $\Sigma \cos \Delta\phi > -0.15$, $m_T < 50$ GeV, 35 GeV $< m_{vis} < 75$ GeV

The acceptance factor A_Z is determined from the generator level Monte Carlo as a ratio of a number of events at the generator level falling into the fiducial region and a number of signal events at the generator level with the $\tau\tau$ invariant mass, before the final state radiation (FSR), in the 66 GeV $< m_{inv} < 116$ GeV mass window. The A_Z factor by construction includes a correction for events migrating from outside the invariant mass window into the fiducial region. The central value for the A_Z factor was calculated using Pythia [22] Monte Carlo generator with the modified LO parton

Table 5.3 Expected number of events for the signal and background processes and the number of events observed in data (N_{obs}) in the $\tau_e\tau_h$ channel

	Number of events in 1.34 fb^{-1}
$\gamma^*/Z \to \ell\ell$	64 ± 4
$W \to \ell\nu$	45 ± 5
$W \to \tau\nu$	18 ± 2
$t\bar{t}$	17 ± 1
Diboson	6 ± 1
Multijet	300 ± 21
Total background	449 ± 22
$\gamma^*/Z \to \tau\tau$	2029 ± 25
N_{obs}	2600

The uncertainties are the statistical uncertainties only

5.5 Methodology for Cross Section Calculation

Table 5.4 Central values for the A_Z acceptance factor and the C_Z correction factor in the $\tau_e \tau_h$ channel

	$\tau_e \tau_h$
A_Z	0.0687 ± 0.0002 (stat.)
C_Z	0.1009 ± 0.0013 (stat.)

distribution function MRSTLO* [23].[3] The Monte Carlo sample contains also low mass γ^*/Z events (the lower bound on the invariant mass is 10 GeV) which might migrate within the fiducial region. The obtained central value in the $\tau_e \tau_h$ final state is mentioned in Table 5.4.

The correction factor C_Z is defined as a ratio of a number of signal events after the full detector simulation which pass all the analysis cuts with all the correction factors (e.g. electron scale factors) applied and a number of events in the fiducial region at the generator level (i.e. the denominator of the C_Z factor is defined in the same way as the A_Z numerator). The C_Z factor includes a correction for migration from outside of the acceptance range by construction. The C_Z factor is calculated using the Alpgen generator with CTEQ6L1 [9] parton distribution function.[4] The derived value is quoted in Table 5.4.

The cross section $\sigma(Z \to \tau\tau)$ defined in Eq. (5.6) is the total inclusive cross section. It is possible to define a so-called fiducial cross section $\sigma^{fid}(Z \to \tau\tau)$ where the knowledge of the acceptance factor A_Z is not required

$$\sigma^{fid}(Z \to \tau\tau) \times BR = \frac{N_{obs} - N_{bkg}}{C_Z \cdot \mathcal{L}}. \tag{5.7}$$

The advantage of the fiducial cross section is that the extrapolation from the fiducial region to the full phase space is not performed. Consequently it is not sensitive to the theoretical uncertainties of the extrapolation model.

5.6 Systematics

Several possible sources of systematic uncertainties that can influence the cross section calculation, uncertainties on the A_Z and C_Z factors as well as on the background estimation, have been studied. The individual sources of systematics in the

---

[3] The Alpgen generator with CTEQ6L1 parton distribution function is used in the fully simulated signal samples throughout the whole analysis and also for the C_Z calculation. The reason why Alpgen is not used also for the A_Z calculation is a problem with the description of the Z boson rapidity when using the CTEQ6L1 PDF set. This problem is not expected to affect the reconstruction level description of event kinematics, but it could affect the extrapolation to the total cross section. Therefore it was decided to use Pythia with MRSTLO* for the A_Z calculation.

[4] The Alpgen generator is a tree-level matrix element calculator for a fixed number of partons. It gives a more precise description for processes with high jet multiplicities compared to generators where the additional jets are produced only during the shower evolution (e.g. Pythia). Furthermore, higher statistics has been available in Alpgen samples than in Pythia.

74   5 $Z \to \tau\tau$ Cross Section Measurement

$\tau_e\tau_h$ channel are discussed one by one in this section. Details about the treatment of systematic uncertainties in the $\tau_\mu\tau_h$ and $\tau_e\tau_\mu$ channels can be found in Ref. [1].

5.6.1 Uncertainties Associated with Electrons

The dominant uncertainty connected to the electron's reconstruction comes from the Monte Carlo simulations of the electron trigger, identification and isolation efficiency. Systematics associated to the electron cleaning and electron energy resolution is also evaluated. The uncertainty associated with the electron energy scale is treated together with the τ energy scale and missing transverse energy uncertainty and is discussed separately in Sect. 5.6.3.

Electron Efficiency

The efficiency of the electron identification, trigger[5] and isolation criteria are measured by means of $W \to e\nu$ and $Z \to ee$ tag and probe methods as described in Chap. 4. The measurements performed on data are compared with the Monte Carlo predictions and so-called scale factors defined as a ratio of efficiency measured in data and in Monte Carlo ($\epsilon_{data}/\epsilon_{MC}$) are derived. These scale factors are used to correct Monte Carlo efficiencies to agree with measurements on data samples. The systematics on the scale factors come mainly from the tag and probe method itself, the derivation of the systematic uncertainties on each factor is also discussed in Chap. 4.

The systematic uncertainty arising from the electron efficiency measurements is evaluated in a conservative way by treating the uncertainties of all scale factors as uncorrelated. The total electron scale factor is defined as a product of the electron reconstruction, identification, trigger and isolation scale factors and its relative uncertainty is evaluated by adding the relative uncertainties of the individual scale factors in quadrature. The uncertainty related to the electron efficiency is calculated by varying the total scale factor by one standard deviation up and down. This approach leads to the uncertainty of 4.8 % on the correction factor C_Z in the $\tau_e\tau_h$ channel. The rather high uncertainty is dominated by the large uncertainty in the identification scale factors for electrons with $E_T < 25\,\text{GeV}$ where a significant part of the signal events occurs.

---

[5] The two components of the combined trigger used in the $\tau_e\tau_h$ channel (electron trigger and hadronic τ trigger) are considered uncorrelated to each other and thus they are measured and applied separately.

Energy Resolution

The Monte Carlo simulations do not reproduce the electron energy resolution as measured in data [17] and a smearing procedure is applied to the simulated samples. The electron resolution uncertainty is evaluated and found to have a very small effect of the order of 0.2 % on the correction factor C_Z.

Electron Cleaning

Systematics related to the use of the object quality check on electrons is considered in the $\tau_e \tau_h$ channel. The systematics arise from the fact that the object quality performed on Monte Carlo simulations is not exactly the same as performed on the data samples, e.g. the dead regions in the electromagnetic calorimeter are not simulated in Monte Carlo, but a correction on the acceptance is applied instead. It is found to be a minor effect of the order of 0.1 % on the correction factor C_Z.

5.6.2 Uncertainties on Hadronic τ Candidates

Two main sources of systematic uncertainty comes from hadronic τ candidates in our analysis: the τ trigger and identification efficiency. The uncertainty on the τ misidentification rate is also evaluated.

Trigger Efficiency

The efficiency of the hadronic τ trigger with respect to the τ candidates passing medium BDT identification are measured in data by means of the $Z \rightarrow \tau\tau$ tag and probe method. The $\tau_e \tau_h$ channel with a single electron trigger is used for this study. The efficiency measured in bins of E_T in the data samples are applied on the Monte Carlo samples as an event weight instead of the τ trigger decision as mentioned in Sect. 5.3.1.

The uncertainty on the measured tau trigger efficiency comes mainly from the background subtraction. The τ trigger weights are varied by one standard deviation up and down to derive the systematic uncertainty associated to the trigger efficiency measurements. This leads to the systematic uncertainty of 4.5 % on the correction factor C_Z.

Identification Efficiency

The identification efficiency of τ candidates has been measured using the tag and probe methods with $Z \rightarrow \tau\tau$ and $W \rightarrow \tau\nu$ events in data collected in 2011 [21].

76   5 $Z \to \tau\tau$ Cross Section Measurement

The average uncertainty on the τ candidates passing medium BDT decision with $E_T > 25$ GeV is 5.1 % in the signal Monte Carlo.

Misidentification Rate

A fake τ candidate can arise from two cases in the $\tau_e\tau_h$ channel, either an electron or a jet can be misidentified as a hadronic τ.

The probability of an electron to be misidentified as a τ candidate has been measured in data using the $Z \to ee$ tag and probe method [21]. Correction factors dependent on the pseudorapidity were evaluated and applied to the Monte Carlo simulations in events with the τ candidate matched to the true electron. These correction factors are varied within their systematic uncertainties, but the effect is found to be negligible in the $\tau_e\tau_h$ channel.

The second case with a jet being misidentified as a hadronic τ is taken into account by normalising all important background sources to data in the specially defined control regions as described in Sect. 5.4. The systematic uncertainty related to this effect is accounted for in the background estimation systematics as described in Sect. 5.6.4.

5.6.3 Energy Scale Uncertainty

The energy scale systematics for electrons, τ candidates and missing transverse energy is considered to be correlated. The uncertainty coming from the energy scale is evaluated accordingly by simultaneously shifting each component up and down by one standard deviation.

The τ energy scale uncertainty is described in details in Ref. [21]. The uncertainty is evaluated from the comparison of the τ candidates' transverse energy distribution for different configurations of the Monte Carlo simulations, e.g. simulations with different showering models or variations of the amount of the dead material in the detector description.

The electron energy scale uncertainty is estimated from the measurements of the $Z \to ee$ events using the precise knowledge of the Z boson mass distribution [17]. Moreover, the energy response was cross-checked in terms of linearity using also $J/\psi \to ee$ and $W \to e\nu$ decays in the central region of the detector.

The systematic uncertainties on the missing transverse energy has been studied in $Z \to \ell\ell$ and $W \to \ell\nu$ events [24]. According to this study, the uncertainty on the E_T^{miss} is evaluated by scaling the energy of all topoclusters in the event up and down by one standard deviation.

The energy scale systematics is found to be dominant in both $\tau_\ell\tau_h$ final states. It is evaluated to be 9.5 % on the correction factor C_Z in the $\tau_e\tau_h$ channel.

5.6 Systematics

5.6.4 Background Estimation

The systematic uncertainty associated to the background estimation for the electroweak processes (W+jets and Z+jets) and multijet background is described in this section.

W+jets and Z+jets Background

The statistical uncertainties of the k_W and k_Z factors used to normalise the electroweak Monte Carlo samples to the data, as described in Sects. 5.4.1 and 5.4.2, are assigned as a systematic uncertainty on the W+jets and Z+jets background estimation. Furthermore, all sources of systematics on the Monte Carlo simulations described above are applied on the W and Z boson Monte Carlo samples and their effects are evaluated. However, the deviations are found to be within the statistical uncertainties of the normalisation factors.

As a cross-check, the normalisation factors are evaluated also using Monte Carlo samples produced with the Pythia generator and the number of W+jets and Z+jets events after the full selection is compared with the results of the default analysis using the Alpgen generator. The numbers of W and Z background events are found to be in a good agreement within the statistical uncertainties. This check supports the assumption that the statistical uncertainty covers the systematic effects related to the τ-jet fake rate in the case of W+jets and Z+jets backgrounds.

Multijet Background

Different sources of systematics enter the total uncertainty on the multijet background estimation. First, the assumption that R_{OSSS} ratio, defined in Eq. (5.5), is independent of the lepton isolation has been checked. The dependence of R_{OSSS} ratio on the track and calorimeter isolation is shown in Fig. 5.11. The maximal deviation from the nominal value of 3 % is found in both lepton-hadron final states. Even though the difference is compatible with the statistical uncertainty of the R_{OSSS} factor, it is still conservatively added to the total uncertainty.

Furthermore, the stability of the R_{OSSS} ratio during the event selection is checked. The maximal difference from the nominal value of 4 % is found in the $\tau_e \tau_h$ channel and is assigned as a systematic uncertainty. The cutflow-dependence of R_{OSSS} is shown in Fig. 5.12.

Another contribution to the multijet background estimation systematics might come from the subtraction of the Monte Carlo simulated events in the control regions B, C and D defined in Sect. 5.4.3. The cross sections of the simulated samples are varied up and down by their uncertainties but this effect is found to be negligible.

All systematic sources mentioned above and the statistical uncertainty of the R_{OSSS} ratio are added in quadrature to obtain the final systematic uncertainty on the multijet background estimation which leads to 1.3 % on the total cross section in the $\tau_e \tau_h$ channel.

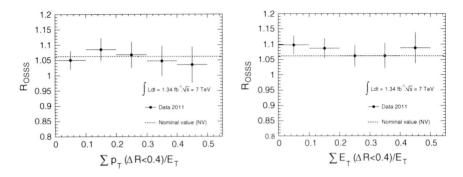

Fig. 5.11 The dependence of the R_{OSSS} ratio on the electron isolation in the $\tau_e\tau_h$ channel. The dependence on the track isolation is shown on the *left* and the calorimeter isolation on the *right*

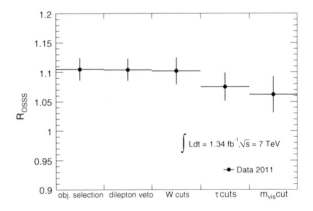

Fig. 5.12 Stability of the R_{OSSS} ratio through the event selection procedure in the $\tau_e\tau_h$ channel. The last point corresponds to the nominal value of R_{OSSS}

5.6.5 Acceptance Factor A_Z Uncertainty

The theoretical uncertainty on the geometric and kinematic acceptance factor A_Z comes mainly from the limited knowledge of the proton parton distribution functions (PDFs) and the uncertainty in the modelling of the Z boson production at the LHC.

Three sources of uncertainties has been considered [1, 2]:

- Uncertainty due to the choice of the PDF set is evaluated as the maximal deviation between the A_Z factor obtained using the default Pythia sample and the values obtained by re-weighting this sample to the CTEQ6.6 and HERAPDF1.0 [25] PDF sets.
- Uncertainty within one PDF set is calculated for the CTEQ6.6 NLO PDF for which 44 PDF error eigenvectors are available [26]. The systematics is obtained by re-weighting the default sample to the relevant CTEQ6.6 error eigenvectors and is compared to the CTEQ6.6 central value.

5.6 Systematics

- Systematic deviation due to the modelling of the parton shower is estimated using the MC@NLO generator interfaced to HERWIG for parton showering.

The uncertainties coming from all three sources are added in quadrature and the total systematic uncertainty of 3.4 % is assigned to the acceptance factor A_Z in the $\tau_e\tau_h$ channel.

5.6.6 Other Sources of Systematic Uncertainty

An uncertainty coming from the background Monte Carlo normalisation is also taken into account. Following Ref. [27], the uncertainty of 5 % on the electroweak background (W boson, Z boson and dibosons) NNLO cross sections is considered. For the $t\bar{t}$ cross section the uncertainty of $+7.0$ %$/-9.6$ % is assumed based on Ref. [28].

The Monte Carlo statistical uncertainty which enters the N_{bkg} calculation is evaluated separately and is found to be 1.4 % effect on the total cross section in the $\tau_e\tau_h$ channel.

The uncertainty on the integrated luminosity is considered to be 3.7 % based on the ATLAS recommendations [29].

The uncertainty associated to the charge misidentification is found to be negligible [2].

5.6.7 Summary of the Systematics

Correlations between the electroweak, $t\bar{t}$ background uncertainties and multijet background uncertainty have to be considered in the evaluation of the systematic uncertainty on the total cross section. The correlation is caused by the fact that the Monte Carlo predictions for W, Z, diboson and $t\bar{t}$ background are subtracted in the control regions used for the multijet background estimation. Therefore the uncertainty on the total cross section from a given source of systematic uncertainty is obtained by recalculating the cross section using at the same time the recalculated C_Z and $(N_{obs} - N_{bkg})$, shifted as indicated for that uncertainty in the corresponding tables. Part of the uncertainties may therefore cancel out.

The effect of the individual systematic sources on the total cross section measurement is presented in Table 5.5. The dominant sources of uncertainties in the $\tau_e\tau_h$ are the energy scale, hadronic τ identification efficiency, followed by the electron efficiency and the τ trigger efficiency uncertainty.

80 5 $Z \to \tau\tau$ Cross Section Measurement

Table 5.5 Relative systematic and statistical uncertainties on the total cross section in the $\tau_e\tau_h$ channel

Uncertainty	$\delta\sigma/\sigma$ (%)
Electron efficiency	5.0
Electron resolution	0.1
Electron cleaning	0.1
τ ID efficiency	5.2
$e - \tau$ misidentification rate	0.2
Energy scale	9.3
Tau trigger efficiency	4.7
W normalization factor	0.04
Z normalization factor	0.05
Multijet estimation	1.3
Background MC normalization	0.2
MC statistics	1.4
A_Z uncertainty	3.4
Total systematic unc.	13.2
Luminosity uncertainty	3.7
Statistical uncertainty	2.4

Table 5.6 The fiducial and total production cross sections for the $Z \to \tau\tau$ process with the $\tau\tau$ invariant mass between 66 and 116 GeV measured in $\tau_e\tau_h$, $\tau_\mu\tau_h$ and $\tau_e\tau_\mu$ channels

Final State	Fiducial cross section $\sigma^{fid}(Z \to \tau\tau) \times BR$
$\tau_\mu\tau_h$	$20.0 \pm 0.3(\text{stat}) \pm 2.0(\text{syst}) \pm 0.7(\text{lumi})$ pb
$\tau_e\tau_h$	$15.9 \pm 0.4(\text{stat}) \pm 2.0(\text{syst}) \pm 0.6(\text{lumi})$ pb
$\tau_e\tau_\mu$	$4.7 \pm 0.2(\text{stat}) \pm 0.4(\text{syst}) \pm 0.2(\text{lumi})$ pb
Final State	Total cross section $\sigma(Z \to \tau\tau)$
$\tau_\mu\tau_h$	$0.91 \pm 0.01(\text{stat}) \pm 0.09(\text{syst}) \pm 0.03(\text{lumi})$ nb
$\tau_e\tau_h$	$1.00 \pm 0.02(\text{stat}) \pm 0.13(\text{syst}) \pm 0.04(\text{lumi})$ nb
$\tau_e\tau_\mu$	$0.96 \pm 0.03(\text{stat}) \pm 0.09(\text{syst}) \pm 0.04(\text{lumi})$ nb

The fiducial cross sections include also the branching fraction of the τ to its decay products

5.7 Final Results

As already mentioned, the $Z \to \tau\tau$ cross section in the $\tau\tau$ invariant mass window 66–116 GeV has been performed in three channels ($\tau_e\tau_h$, $\tau_\mu\tau_h$ and $\tau_e\tau_\mu$) with 2011 data samples. The measurement in the $\tau_e\tau_h$ has been described in details in this chapter whereas more details about the other two channels can be found in Ref. [1]. The measured cross sections, the fiducial cross section and the total cross section with their uncertainties, in all three final states are summarized in Table 5.6. The total cross section has been corrected for the $\tau \to \ell\nu\nu$ and $\tau \to \tau_h\nu$ branching ratios according to Ref. [4]: 0.2313 ± 0.0009 in the $\tau_e\tau_h$ channel, 0.2250 ± 0.0009 in the $\tau_\mu\tau_h$ channel and 0.0620 ± 0.0002 in the $\tau_e\tau_\mu$ channel.

5.7 Final Results

Table 5.7 Assumed correlations between individual channels. The uncertainties are considered either 100 % correlated ($\checkmark$) or fully uncorrelated (–)

Uncertainty	$\tau_e\tau_h$	$\tau_\mu\tau_h$	$\tau_e\tau_\mu$
Muon efficiency	–	$\checkmark$	$\checkmark$
Electron efficiency	$\checkmark$	–	$\checkmark$
Muon resolution	–	$\checkmark$	$\checkmark$
Electron resolution	$\checkmark$	–	$\checkmark$
Jet resolution	–	–	–
τ ID efficiency	$\checkmark$	$\checkmark$	–
$e - \tau$ misidentification rate	–	–	–
Energy scale	$\checkmark$	$\checkmark$	$\checkmark$
τ trigger efficiency	–	–	–
W normalization factor	–	–	–
Z normalization factor	–	–	–
Multijet estimation	$\checkmark$	$\checkmark$	$\checkmark$
Background MC normalization	$\checkmark$	$\checkmark$	$\checkmark$
MC statistics	–	–	–
A_Z uncertainty	$\checkmark$	$\checkmark$	$\checkmark$
Luminosity uncertainty	$\checkmark$	$\checkmark$	$\checkmark$

The measured cross sections in the three final states has been combined by means of the Best Linear Unbiased Estimate (BLUE) [30, 31]. The BLUE method gives the best estimate of the combined cross section by a linear combination of the individual measurements. A covariance matrix is built from the statistical and systematic uncertainties for each individual measurement and the correlations of the uncertainties from each channel are accounted for in the BLUE method. The considered correlations are summarized in Table 5.7. The systematic uncertainties for the same physics objects are considered fully correlated while there is no correlation assumed between uncertainties related to different objects.

Most of the dominant uncertainties are correlated across all three channels and they are much larger than any uncorrelated systematics. The BLUE method can lead to a combination which is outside the range spanned by the individual measurements in case of the large positive correlations [30]. In order to avoid this behaviour, the largest systematics fully correlated across all three channels (energy scale, luminosity and acceptance uncertainties) are excluded from the BLUE calculation. These uncertainties are not considered for the mean value of the combination, but they are added to the final uncertainty using the standard error propagation on the linear combination of the individual cross sections with weights derived by the BLUE method. This approach leads to a relatively larger total uncertainty on the combination since the three large uncertainties has not been included in the combination.

Following the described procedure, the $Z \to \tau\tau$ combined cross section with the $\tau\tau$ invariant mass within $66 < m_{\tau\tau} < 116$ GeV of

$$\sigma(Z \to \tau\tau) = 0.92 \pm 0.02(\text{stat}) \pm 0.08(\text{syst}) \pm 0.03(\text{lumi}) \text{ nb} \qquad (5.8)$$

is calculated with corresponding weights of 0.758 for $\tau_\mu\tau_{\rm h}$, -0.130 for $\tau_e\tau_{\rm h}$ and 0.372 for $\tau_e\tau_\mu$. A $\chi^2/$NDF of 1.24/2 is obtained. The combined cross section agrees well with the NNL0 theoretical expectations of 0.96 ± 0.05 nb [10–12].

References

1. The ATLAS Collaboration, G. Aad et al., $Z \to \tau\tau$ cross section measurement in proton-proton collisions at 7 TeV with the ATLAS experiment, ATL-CONF-2012-006, Feb 2012
2. The ATLAS Collaboration, G. Aad et al., Measurement of the Z to tau tau cross section with the ATLAS detector. Phys. Rev. D **84**, 112006 (2011)
3. The CMS Collaboration, Measurements of inclusive W and Z cross section via decays to Tau Pairs in pp collisions at $\sqrt{s} = 7$ TeV. JHEP **08**, 117 (2011)
4. Particle Data Group Collaboration, K. Nakamura et al., Review of particle physics. J. Phys. **G37**, 075021 (2010)
5. The GEANT4 Collaboration, S. Agostinelli et al., GEANT4: a simulation toolkit. Nucl. Instrum. Meth. **A506**, 250 (2003)
6. M.L. Mangano, F. Piccinini, A.D. Polosa, M. Moretti, R. Pittau, ALPGEN, a generator for hard multiparton processes in hadronic collisions. JHEP **07**, 001 (2003)
7. G. Corcella et al., HERWIG 6: an event generator for hadron emission reactions with interfering gluons (including supersymmetric processes). JHEP **01**, 010 (2001)
8. J.M. Butterworth, J.R. Forshaw, M.H.Seymour, Multiparton interactions in photoproduction at HERA. Z.Phys. C **72**, 637–646 (1996)
9. J. Pumplin, D.R. Stump, J. Huston, H.-L. Lai, P. Nadolsky, W.-K. Tung, New generation of parton distributions with uncertainties from global QCD analysis. JHEP **07**, 012 (2002)
10. K. Melnikov, F. Petriello, Electroweak gauge boson production at hadron colliders through $O(\alpha(s)^2)$. Phys. Rev. **D74**, 114017 (2006)
11. F.P.R. Gavin, Y.Li et al., FEWZ 2.0: a code for hadronic Z production at next-to-next-to- leading order. CPC **182**, 2388 (2011)
12. S. Catani, L. Cieri, G. Ferrera, D. de Florian, M. Grazzini, Vector boson production at hadron colliders: a fully exclusive QCD calculation at NNLO. Phys. Rev. Lett. **103**, 082001 (2009)
13. S. Frixione, B. Webber, Matching NLO QCD computations and parton shower simulations. JHEP **0206**, 029 (2002)
14. S. Jadach, Z. Was, R. Decker, J.H.Kühn, The τ decay library TAUOLA, version 2.4. Comput. Phys. Commun. **76**, 361–380 (1993)
15. P. Golonka, Z. Was, PHOTOS Monte Carlo: a precision tool for QED corrections in Z and W decays. EPJC **45**, 97–107 (2006)
16. The ATLAS Collaboration, Luminosity public results (2011), https://twiki.cern.ch/twiki/bin/view/AtlasPublic/LuminosityPublicResults
17. The ATLAS Collaboration, G. Aad et al., Electron performance measurements with the ATLAS detector using the 2010 LHC proton-proton collision data. EPJC **72**(2012), 1909
18. The ATLAS Collaboration, Aad et al., Measurement of the inclusive $W^\pm$ and Z/γ^* cross sections in the electron and muon decay channels in pp collisions at $\sqrt{s} = 7$ TeV with the ATLAS detector. Phys. Rev. **D85**, 072004 (2012)
19. M. Cacciari, G.P. Salam, Dispelling the N_3 myth for the kt jet-finder. Phys. Lett. B **641**, 57 (2006)
20. M. Cacciari, G.P. Salam, G. Soyez, http://fastjet.fr/
21. The ATLAS Collaboration, G. Aad et al., Performance of the reconstruction and identification of hadronic tau decays with ATLAS, ATLAS-CONF-2011-152, Nov 2011
22. T. Sjostrand, S. Mrenna, P. Skands, PYTHIA 6.4 physics and manual. JHEP **05**, 026 (2006)
23. A. Sherstnev, R.S. Thorne, Parton distributions for LO generators. EPJC **55**, 553 (2008)

References

24. The ATLAS Collaboration, G. Aad et al., Performance of missing transverse momentum reconstruction in proton-proton collisions at 7 TeV with ATLAS. EPJC **72**, 1844 (2012)
25. H1, ZEUS Collaboration, Combined measurement and QCD analysis of the inclusive e p scattering cross sections at HERA. JHEP **1001**, 109 (2010)
26. P.M. Nadolsky et al., Implications of CTEQ global analysis for collider observables. Phys. Rev. **D78**, 013004 (2008)
27. The ATLAS Collaboration, G. Aad et al., Measurement of the $W \rightarrow \ell\nu$ and $Z/\gamma^* \rightarrow \ell\ell$ production cross sections in proton–proton collisions at $\sqrt{s} = 7$ TeV with the ATLAS detector. JHEP **12**, 60 (2010)
28. The ATLAS Collaboration, G. Aad et al., Measurement of the top-quark pair production cross-section in pp collisions at $\sqrt{s} = 7$ TeV in dilepton nal states with ATLAS, ATLAS-CONF-2011-100, Oct 2011
29. The ATLAS Collaboration, G. Aad et al., Luminosity determination in pp collisions at $\sqrt{s} = 7$ TeV using the ATLAS detector in 2011, ATLAS-CONF-2011-116, Aug 2011
30. L. Lyons, D. Gibaut, P. Clifford, How to combine correlated estimates of a single physical quantity. Nucl. Instrum. Meth. **A270**, 110 (1988)
31. A. Valassi, Combining correlated measurements of several different physical quantities. Nucl. Instrum. Meth. **A500**, 391–405 (2003)

Chapter 6
Conclusions

The Monte Carlo simulations of the Tile Calorimeter have been described, especially the performance of the electronic noise and the pile-up simulations. The double gaussian noise shape has been implemented in the simulations and it leads to a good agreement of the cell energy spectrum with data. However, the electronic noise description does not include all features observed in data, namely correlations between individual channels. The dependence of the pile-up contribution on the average number of minimum bias collisions per bunch crossing and on the bunch separation has been shown. The pile-up contribution is evaluated by means of the cell energy spread in bins of $|\eta|$ and radial sample for the considered pile-up conditions. The double gaussian parameters as well as pile-up constants are inserted in the database and are used to define 2σ and 4σ limits in the clustering algorithm.

The efficiencies of the electron identification, trigger and isolation cuts have been measured by means of the $W \rightarrow e\nu$ and $Z \rightarrow ee$ tag and probe methods. The scale factors, ratios of the efficiencies measured in data and in Monte Carlo simulations, with their statistical and systematic uncertainties have been derived in bins of electron pseudorapidity and transverse energy. These factors have been used in the $Z \rightarrow \tau\tau$ cross section measurement to correct the Monte Carlo predictions.

The measurement of the $Z \rightarrow \tau\tau$ cross section with the electron and the hadronic τ in the final state, so-called $\tau_e\tau_h$ channel, has been discussed in details. Data collected during 2011 data taking period corresponding to 1.34 fb^{-1} has been used in the analysis. The methodology of the cross section measurement within the $\tau\tau$ invariant mass range 66–116 GeV, the calculation of the nominal value and the evaluation of the systematic uncertainties have been described. The main sources of the systematic uncertainty in the $\tau_e\tau_h$ channel are the energy scale, hadronic τ identification and trigger efficiency and electron efficiency. The cross section measurement has been performed also in the $\tau_\mu\tau_h$ and $\tau_e\tau_\mu$ final states. These measurements are not discussed in details, but they are documented in Ref. [1]. The final total cross section is derived as a combination of the individual measurements in the $\tau_e\tau_h$, $\tau_\mu\tau_h$ and $\tau_e\tau_\mu$ channels by means of the BLUE method. The combined cross section of the $Z \rightarrow \tau\tau$ process with the $\tau\tau$ invariant mass between 66 and 116 GeV is evaluated

J. Nováková, *Standard Model Measurements with the ATLAS Detector*,
Springer Theses, DOI: 10.1007/978-3-319-00810-3_6,
© Springer International Publishing Switzerland 2014

as $0.92 \pm 0.02(\text{stat}) \pm 0.08(\text{syst}) \pm 0.03(\text{lumi})$ nb which is in a good agreement with the NNL0 theoretical expectations of 0.96 ± 0.05 nb.

Reference

1. The ATLAS Collaboration, G. Aad et al., $Z \rightarrow \tau\tau$ cross section measurement in proton–proton collisions at 7 TeV with the ATLAS experiment, ATL-CONF-2012-006, Feb 2012

Appendix A
Single Electron Trigger Scale Factors

The methodology and measurement of single electron trigger efficiencies and scale factors by means of the $W \to e\nu$ tag and probe method is described in details in Sect. 4.4. The results, the efficiencies and the scale factors in bins of $E_T \times |\eta|$, for the trigger EF_e15_medium[1] are summarised in Table A.1.

Table A.1 Single electron trigger efficiencies in % for data (ϵ_{data}) and Monte Carlo simulations (ϵ_{MC}) together with scale factors (SF) for trigger EF_e15_medium in bins of $E_T \times |\eta|$

		$|\eta| \in (0.0, 0.8)$	$|\eta| \in (0.8, 1.37)$	$|\eta| \in (1.52, 2.47)$
$E_T \in (17, 20)$ GeV	ϵ_{data}	98.17 ± 0.21	96.72 ± 0.36	95.95 ± 0.32
	ϵ_{MC}	99.26 ± 0.18	99.45 ± 0.19	97.02 ± 0.33
	SF	0.9889 ± 0.0027	0.9725 ± 0.0040	0.9890 ± 0.0048
$E_T \in (20, 30)$ GeV	ϵ_{data}	99.16 ± 0.03	98.60 ± 0.05	97.51 ± 0.05
	ϵ_{MC}	99.79 ± 0.02	99.64 ± 0.04	98.31 ± 0.06
	SF	0.9937 ± 0.0004	0.9896 ± 0.0006	0.9918 ± 0.0008
$E_T > 30$ GeV	ϵ_{data}	99.37 ± 0.01	99.40 ± 0.02	97.49 ± 0.03
	ϵ_{MC}	99.71 ± 0.01	99.65 ± 0.02	98.54 ± 0.04
	SF	0.9965 ± 0.0002	0.9975 ± 0.0003	0.9894 ± 0.0004

The data sample corresponds to 2.1 fb^{-1}. The total uncertainty (statistical and systematic errors summed in quadrature) is quoted

---

[1] The trigger EF_e15_medium searches for an electron object with transverse energy larger than 15 GeV at the event filter (EF) level.

J. Nováková, *Standard Model Measurements with the ATLAS Detector*,
Springer Theses, DOI: 10.1007/978-3-319-00810-3,
© Springer International Publishing Switzerland 2014

Appendix B
Electron Isolation Scale Factors

The methodology and measurement of electron isolation criteria efficiencies and scale factors by means of the $Z \to ee$ tag and probe method is described in details in Sect. 4.5. The results, the efficiencies and the scale factors in bins of $E_T \times \eta$, for the electron isolation cuts used in the $Z \to \tau\tau$ analysis ($p_T\text{Cone}40/E_T < 0.06$ and $E_T\text{Cone}40/E_T < 0.1$) are summarised in Tables B.1 and B.2.

J. Nováková, *Standard Model Measurements with the ATLAS Detector*,
Springer Theses, DOI: 10.1007/978-3-319-00810-3,
© Springer International Publishing Switzerland 2014

Table B.1 First part ($E_T < 40$ GeV): electron isolation efficiencies in % for data (ϵ_{data}) and Monte Carlo simulations (ϵ_{MC}) together with scale factors (SF) for the considered isolation criteria ($p_T\text{Cone}40/E_T < 0.06$ and $E_T\text{Cone}40/E_T < 0.1$) in bins of $E_T \times \eta$

		$E_T \in (17, 30)$	$E_T \in (30, 40)$
$\eta \in (-2.47, -2.01)$	ϵ_{data}	68.8 ± 0.7	75.8 ± 0.5
	ϵ_{MC}	75.9 ± 0.5	82.6 ± 0.4
	SF	$0.907 \pm 0.011 \pm 0.008$	$0.917 \pm 0.007 \pm 0.007$
$\eta \in (-2.01, -1.52)$	ϵ_{data}	65.5 ± 0.7	73.7 ± 0.4
	ϵ_{MC}	70.4 ± 0.5	78.2 ± 0.3
	SF	$0.930 \pm 0.011 \pm 0.006$	$0.942 \pm 0.007 \pm 0.003$
$\eta \in (-1.37, -0.8)$	ϵ_{data}	64.8 ± 0.6	74.2 ± 0.3
	ϵ_{MC}	70.8 ± 0.5	80.9 ± 0.2
	SF	$0.915 \pm 0.010 \pm 0.010$	$0.918 \pm 0.005 \pm 0.004$
$\eta \in (-0.8, -0.1)$	ϵ_{data}	72.1 ± 0.4	82.5 ± 0.2
	ϵ_{MC}	77.4 ± 0.3	86.4 ± 0.2
	SF	$0.930 \pm 0.007 \pm 0.004$	$0.954 \pm 0.003 \pm 0.001$
$\eta \in (-0.1, 0.1)$	ϵ_{data}	73.3 ± 0.8	83.8 ± 0.5
	ϵ_{MC}	78.5 ± 0.6	86.6 ± 0.3
	SF	$0.934 \pm 0.013 \pm 0.008$	$0.968 \pm 0.007 \pm 0.002$
$\eta \in (0.1, 0.8)$	ϵ_{data}	72.1 ± 0.4	83.0 ± 0.2
	ϵ_{MC}	77.4 ± 0.3	86.6 ± 0.2
	SF	$0.932 \pm 0.007 \pm 0.006$	$0.958 \pm 0.003 \pm 0.002$
$\eta \in (0.8, 1.37)$	ϵ_{data}	64.2 ± 0.6	74.9 ± 0.3
	ϵ_{MC}	72.2 ± 0.5	81.3 ± 0.2
	SF	$0.889 \pm 0.010 \pm 0.011$	$0.921 \pm 0.005 \pm 0.002$
$\eta \in (1.52, 2.01)$	ϵ_{data}	66.8 ± 0.6	74.9 ± 0.4
	ϵ_{MC}	70.3 ± 0.5	78.3 ± 0.3
	SF	$0.950 \pm 0.012 \pm 0.008$	$0.956 \pm 0.007 \pm 0.005$
$\eta \in (2.01, 2.47)$	ϵ_{data}	69.1 ± 0.7	76.5 ± 0.5
	ϵ_{MC}	76.0 ± 0.5	82.8 ± 0.4
	SF	$0.909 \pm 0.011 \pm 0.015$	$0.924 \pm 0.007 \pm 0.002$

The data sample corresponds to 1.3 fb^{-1}. The statistical uncertainties are quoted for the efficiencies, while both statistical (first error) and systematic uncertainties (second error) are given for the scale factors

Appendix B Electron Isolation Scale Factors   91

Table B.2 Second part ($E_T > 40$ GeV): electron isolation efficiencies in % for data (ϵ_{data}) and Monte Carlo simulations (ϵ_{MC}) together with scale factors (SF) for the considered isolation criteria (p_TCone40/$E_T < 0.06$ and E_TCone40/$E_T < 0.1$) in bins of $E_T \times \eta$

		$E_T \in (40, 50)$	$E_T > 50$
$\eta \in (-2.47, -2.01)$	ϵ_{data}	83.1 ± 0.4	87.5 ± 0.7
	ϵ_{MC}	88.2 ± 0.3	92.1 ± 0.5
	SF	$0.942 \pm 0.006 \pm 0.002$	$0.951 \pm 0.009 \pm 0.005$
$\eta \in (-1.52, -2.01)$	ϵ_{data}	83.9 ± 0.3	89.4 ± 0.5
	ϵ_{MC}	86.5 ± 0.3	91.3 ± 0.4
	SF	$0.970 \pm 0.005 \pm 0.001$	$0.979 \pm 0.007 \pm 0.004$
$\eta \in (-1.37, -0.8)$	ϵ_{data}	83.3 ± 0.2	89.7 ± 0.4
	ϵ_{MC}	89.0 ± 0.2	93.6 ± 0.2
	SF	$0.936 \pm 0.003 \pm 0.004$	$0.958 \pm 0.004 \pm 0.001$
$\eta \in (-0.8, -0.1)$	ϵ_{data}	89.5 ± 0.2	94.3 ± 0.2
	ϵ_{MC}	92.7 ± 0.1	95.9 ± 0.2
	SF	$0.966 \pm 0.002 \pm 0.001$	$0.984 \pm 0.003 \pm 0.002$
$\eta \in (-0.1, 0.1)$	ϵ_{data}	90.2 ± 0.3	94.4 ± 0.5
	ϵ_{MC}	93.1 ± 0.2	96.3 ± 0.3
	SF	$0.969 \pm 0.004 \pm 0.001$	$0.980 \pm 0.006 \pm 0.005$
$\eta \in (0.1, 0.8)$	ϵ_{data}	89.8 ± 0.2	94.2 ± 0.2
	ϵ_{MC}	92.7 ± 0.1	96.2 ± 0.1
	SF	$0.969 \pm 0.002 \pm 0.001$	$0.980 \pm 0.003 \pm 0.001$
$\eta \in (0.8, 1.37)$	ϵ_{data}	84.0 ± 0.3	90.5 ± 0.4
	ϵ_{MC}	88.8 ± 0.2	94.0 ± 0.2
	SF	$0.946 \pm 0.004 \pm 0.001$	$0.963 \pm 0.005 \pm 0.006$
$\eta \in (1.52, 2.01)$	ϵ_{data}	84.0 ± 0.3	90.5 ± 0.5
	ϵ_{MC}	86.8 ± 0.3	91.6 ± 0.4
	SF	$0.968 \pm 0.005 \pm 0.001$	$0.988 \pm 0.007 \pm 0.001$
$\eta \in (2.01, 2.47)$	ϵ_{data}	82.6 ± 0.4	88.5 ± 0.7
	ϵ_{MC}	88.7 ± 0.3	92.8 ± 0.5
	SF	$0.930 \pm 0.006 \pm 0.004$	$0.954 \pm 0.008 \pm 0.006$

The data sample corresponds to 1.3 fb^{-1}. The statistical uncertainties are quoted for the efficiencies, while both statistical (first error) and systematic uncertainties (second error) are given for the scale factors

Printed by Publishers' Graphics LLC
DBT130903.15.16.80